Temperature-Programmed Reduction for Solid Materials Characterization

CHEMICAL INDUSTRIES

A Series of Reference Books and Textbooks

Consulting Editor
HEINZ HEINEMANN
Heinz Heinemann, Inc.,
Berkeley, California

Temperature-Programmed Reduction for Solid Materials Characterization

Alan Jones

Explosion and Flames Laboratory
Buxton, England

Brian McNicol

Thornton Research Center
Chester, England

CRC Press
Taylor & Francis Group
Boca Raton London New York

CRC Press is an imprint of the
Taylor & Francis Group, an **informa** business

First published 1986 by Marcel Dekker

Published 2019 by CRC Press
Taylor & Francis Group
6000 Broken Sound Parkway NW, Suite 300
Boca Raton, FL 33487-2742

First issued in paperback 2019

No claim to original U.S. Government works

ISBN 13: 978-0-367-45159-2 (pbk)
ISBN 13: 978-0-8247-7583-4 (hbk)

Visit the Taylor & Francis Web site at
http://www.taylorandfrancis.com

and the CRC Press Web site at
http://www.crcpress.com

Library of Congress Cataloging-in-Publication Data

Jones, Alan
   Temperature-programmed reduction for solid materials characterization.

   (Chemical industries ; v. 24)
   Bibliography: p.
   Includes index.
   1. Reduction, Chemical. 2. Materials--Analysis.
I. McNicol, Brian D. II. Title. III. Series.
QD63.R4J66   1986   545'.23   86-6291
ISBN 0-8247-7583-X

PREFACE

Temperature-programmed reduction (TPR) is a relatively new technique used for the characterization of solid materials. Since the first publication in this area (1975), many papers have appeared in the literature that have shown its applicability in a variety of scientific fields, most particularly in characterization of catalysts. In 1982 the first review on the subject of TPR appeared in *Catalysis Reviews*.

The technique, based on the reducibility of species in solids, enables one to obtain information not only of a purely analytical nature but also, and more importantly, about the condition of species present in and on solids. At the same time, the technique is relatively simple in concept and application. It involves very low capital investment relative to more sophisticated techniques used in the above area.

As a result, TPR and related techniques are now becoming commonly used weapons in the scientist's armory for the characterization of solids. This applies across the whole spectrum of academic and industrial research. It is thus timely that the first book should appear that represents not only the state of the art of the technique, but also recent developments and future trends.

The book is intended as an informative guide to both experts and initiates in the field. The principles and theory of isothermal and dynamic reducibility (TPR) are discussed in full detail, together with a clear description of the parametric sensitivity of the technique (Chapter 2). Chapter 3 describes the instrumentation that has been used for the measurement of TPR. Chapter 4 fully discusses the areas where the technique has found application in such a way that the reader should gain ready knowledge of the current state of the art and the possible advantages and disadvantages of the technique in comparison with other techniques. Finally, in Chapter 5 the developing trends in temperature programming are discussed. Here we are concerned with not only TPR but temperature-programmed reactions in general. This chapter attempts to point the direction in which temperature-programming techniques for the characterization of materials are heading.

This volume is aimed at a wide audience, including academics (undergraduate, postgraduate and university staff), research scientists, and technologists in industry and research institutes concerned with the manifold problems involving characterization of solids in such fields as heterogeneous catalysis, materials science/metallurgy, and analytical chemistry.

The terminology used in this and related fields has led to some confusion; in particular, the acronym TPR has been used to refer to both temperature-programmed reduction and temperature-programmed reaction. We would recommend and have used throughout this work the abbreviation TPR to refer to temperature-programmed reduction. Where reference is made to temperature-programmed reaction the expression is written in full. Abbreviations relating to specific temperature-programmed reactions present, so far, no ambiguity and we have recommended the following:

Temperature-programmed oxidation (TPO)

Temperature-programmed sulfidation (TPS)

Temperature-programmed carburization (TPC)

Temperature-programmed methanation (TPM)

We would like to thank Drs. S. J. Gentry and N. W. Hurst of the Health and Safety Executive, who have done much pioneering in this field and did much of the groundwork necessary for the preparation of this work. We have drawn freely from the original review article, of which they were coauthors, in the preparation of this book. We would also like to thank Mr. I. Kitchener of the London Research Station of British Gas for providing us with advice on optimizing conditions for TPR. Finally, thanks are due to all the scientists who provided details of their work for inclusion in the book and permission to reproduce diagrams.

Alan Jones

Brian McNicol

CONTENTS

Temperature-Programmed Reduction
for Solid Materials
Characterization

1

Introduction

I. BACKGROUND

Temperature-programmed reduction (commonly abbreviated to TPR) is a technique used for the chemical characterization of solids. The essence of the technique is the reduction of a solid by a gas at the same time that the temperature of the system is changed in a predetermined way. Chemical information is derived from a record of the analysis of the gaseous products.

In the most commonly encountered apparatus, the solid is reduced by flowing hydrogen, the concentration of which is monitored downstream of the reactor. Provided that reduction has taken place over the temperature excursion of the reactor, the analysis record is simply the hydrogen consumption and is usually displayed as a function of temperature of the reactor.

A typical reduction profile, as shown in Fig. 1, consists of a series of peaks. Each peak represents a distinct reduction process involving a particular chemical component of the solid. The position of a peak in the profile is determined by the chemical nature and environment of the chemical component, and the area of the peak reflects the concentration of that component present in the solid.

TPR is a relatively new technique for characterizing solids. It is highly sensitive and does not depend on any specific property of the solid under investigation other than its reducibility. The technique was first investigated about 13 years ago by Jenkins [1] during a study of catalysts using the temperature-programmed desorption technique pioneered by Cvetanovic and Amenomiya [2]. Jenkins realized that valuable information could be obtained from the reduction pretreatment stage of this technique if the reduction was also performed in the temperature-programmed mode. A series of experiments provided data which, together with information from other techniques, enabled the catalysts to be fully characterized.

The technique was soon exploited by others, and a notable early study [3] was made using the technique to characterize a series of solid catalysts of noble metals supported on refractory oxides. Over the last decade the technique has been exploited and developed considerably by workers around the world to the extent that TPR now has a place alongside other techniques for the characterization of solids, and in 1982 the first review article on the subject was published [4].

The technique can be regarded as the principal member of a family of temperature-programmed reaction techniques. Its

status as principal member is only due to the fact that the reduction reaction has attracted the most attention from research-
ers. In principle, valuable information can be obtained by inves-
tigating many different types of chemical reactions in a temperature-programmed mode. Using the same or similar apparatus,
temperature-programmed oxidations, methanations, sulfidations,
and carburizations have been reported.

On the basis of the type of information it can provide, the
TPR technique can be classified with the spectroscopic and X-
ray techniques that have been traditionally and indeed are currently used to characterize solids. Almost all of these techniques
impose severe requirements on the solid or on the conditions
under which it can be investigated. These requirements neces-
sarily limit the information that can be obtained and the range

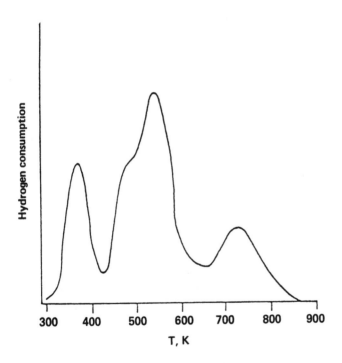

Fig. 1 A typical reduction profile.

of solids that can be investigated. The TPR technique, while capable of providing chemical information similar to that from the other techniques, has, in general, a much wider range of application. Even the condition that the solid must be reducible is not mandatory, since valuable information has been obtained on *reduced* solids using TPR by first mildly oxidizing the solids and then subjecting them to the TPR experiment.

In another sense, and certainly on the grounds of similarity of experimental procedure, TPR can be classified with a broad range of techniques, known as thermoanalytical techniques, that are used to generate physical and chemical information on solids [5]. In general, these techniques depend on the measurement of a parameter of a physical or chemical property of a solid as the temperature of the solid is varied, as for TPR, in a predetermined manner. Figure 2 presents a classification of the principal techniques. These techniques provide information on a wide range of phenomena from the purely physical to the totally chemical [5]. Temperature-programmed desorption (TPD) is included in Fig. 2 under techniques dependent on the evolution of "volatiles" since there are broad experimental similarities with evolved gas analysis (EGA) and both techniques can give information on the physical desorption of gases as well as on gases evolved by chemical mechanisms. Temperature-programmed reduction, along with the other temperature programmed reaction techniques, is thus identified as a more chemically based technique concerned with analysis of gases from purely chemical processes.

Temperature-programmed reduction profiles of complex solids can show a great deal of detail, reflecting complex reduction processes. The use of such profiles as "fingerprints" to provide a rapid assessment of the correctness of the composition of a solid has proved useful in many applications. In particular, the fingerprint temperature reduction profile has found application as a tool for quality control in the preparation of industrial catalysts.

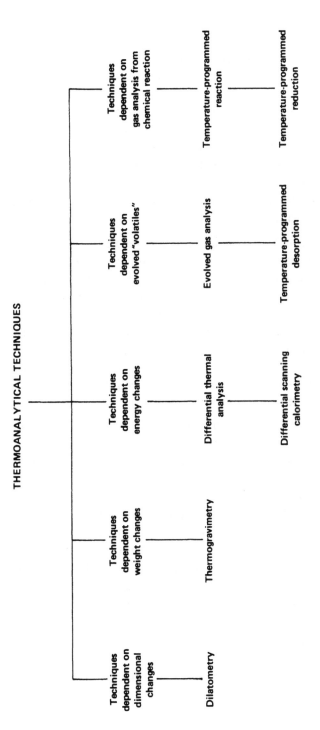

Fig. 2 A classification of thermoanalytical techniques.

5

Although many analytical techniques are also capable of providing similar fingerprint analysis, for solid catalysts temperature-programmed reaction techniques have the advantage that the fingerprint can more truly reflect the chemical and structural features essential for its use because the analytical reaction and reaction conditions can be made closely similar to the conditions experienced by the catalyst in its industrial use. Thus, a temperature-programmed reduction profile can provide a fingerprint reflecting the essential features of a reduction catalyst, just as a temperature-programmed methanation profile can for a methanation catalyst.

In this role in particular, three other factors have contributed to the success of the TPR technique. First, the sensitivity is such that a reduction profile can be obtained from the consumption of as little as 1 μm of hydrogen, requiring only milligrams of sample. Second, the apparatus, consisting of simple temperature programmers and thermal conductivity detectors, is inexpensive when compared with that required for the X-ray and spectroscopic techniques. Third, the relative robustness of the apparatus and its minimal maintenance requirements make it an ideal tool for industrial application.

The problems facing the science of the characterization of solids include the identification of low concentrations of impurities, or dopants; alloy formation between metals; other interactions between solids (e.g., between a metal and an oxide); particle sizes of components; and the influence of thermal treatments. Experience in the catalyst characterization area has shown that TPR is well suited to the study of such phenomena. Particular advantages have been the few, if any, limitations imposed on materials, the sensitivity of the technique and the ability, for a number of catalysts, to study phenomena in realistic conditions hitherto not possible with other techniques.

II. APPLICATIONS

Thus, over recent years TPR has been extensively applied to the characterization of solid catalysts, and valuable information has been obtained that has led to a greater understanding of the role of the catalyst. In the area of petroleum processing it has been applied to the establishment of the condition of the active metal component, e.g., in Pt/Re reforming catalysts there has been considerable debate regarding the nature of the Re component. TPR has shown that in commercial reforming catalysts under realistic working conditions the Re is reduced to the zero valent state and that alloying between Re and Pt takes place during the actual reduction step of the catalyst preparation. Similarly, TPR work has shown that in other bimetallic reforming catalysts, such as Pt/Ge and Pt/Sn, the nonnoble metal component is not completely reduced. Part is reduced to the zero valent state, with the remainder participating in an interaction with the support. The concentrations of metal are so low ($<1\%$ W/W) and the particle sizes so small (<1 nm) that other techniques have not been able to provide information because of lack of sensitivity.

Other problems successfully tackled by TPR include the characterization of hydroprocessing catalysts where the extent of sulfiding of Mo and W/Al_2O_3 catalysts has been determined; the characterization of Fischer Tropsch catalysts, where reducibility of Ni/SiO_2 was established and alloy formation between Fe and Ni was shown; and the characterization of Rhodium/polyphosphine catalysts, where reducibility of Rh was established.

The interactions that can occur between metals and the support have assumed greater importance over recent years as different techniques have revealed that such phenomena exist. TPR has shown the existence of such an interaction in Pt/Al_2O_3 catalysts, which explained why such catalysts have such a stable dispersion of low-particle-size metal. In Rh/Al_2O_3 exhaust emission catalysts, the dissolution of Rh into Al_2O_3 at high temperatures was established by TPR and explained why at such temperatures the catalyst is no longer available for reaction.

The application of TPR has also thrown much light on the processes involved in the regeneration of industrial catalysts. Particular success has been achieved in the characterization of an important class of compounds, the molecular sieve zeolites.

These compounds have found wide industrial use as catalysts and
in a number of separation processes. Their structure is essentially an aluminosilicate matrix into which are dispersed metal
cations occupying distinct chemical sites. Because of the small
crystallite size, characterization is difficult using conventional X-
ray diffraction techniques and powder X-ray diffraction gives
only limited information on cation positions. Using TPR the
degree of reduction has been determined, and from the reduction profiles the specific location of the cations, so vital to the
success of the zeolite as a catalyst or separating agent, has been
established.

Similarly, in the field of electrochemistry our understanding of the condition of electrocatalytically active electrodes is
important in the development of new electrode materials and
better electrochemical processes. Recently TPR has been successfully applied to the characterization of carbon-supported
noble metal electrocatalysts for development of acid electrolyte
methanol-air fuel cells.

All of the advances alluded to above are fairly recent, and
now that the ability of the technique has been demonstrated the
possibilities for further advances are enormous. However, quite
apart from the above mainly catalysis oriented work, there are
opportunities for application of the technique in general to the
characterization of solids.

From the chemical analysis point of view, the technique has
the advantages of being sensitive and nondestructive—in the sense
that the solid does not have to be dissolved—and is particularly
strong on the identification of the valence states of metallic
components.

Since much TPR work has been concerned with the reduction of oxide species and recent work has dealt with the characterization of sulfides, applications are apparent in corrosion science, where the identification and quantification of the components of metallic oxide and sulfide deposits are important.

There is an increased interest in coal chemistry arising from the increasing importance of coal, in certain areas in the world, as an energy source and as a feedstock for the production of chemicals. TPR is particularly suited to the characterization of coal ashes and should be capable of elucidating the nature of the reducible species to a greater extent than is possible with existing techniques. In this respect a considerable amount of work has already been done on the TPR of methanation catalysts linked to coal gasification processes.

REFERENCES

1. J. W. Jenkins, Gordon Research Conference on Catalysis, Colby-Sawyer College, New London, New Hampshire (1975)

2. R. J. Cvetanovic and Y. Amenomiya, *Catal. Rev.*, *6* (1): 21 (1972)

3. S. D. Robertson, B. D. McNicol, J. H. De Baas, S. C. Kloet, and J. W. Jenkins, *J. Catal.*, *37*: 424 (1975)

4. N. W. Hurst, S. J. Gentry, A. Jones, and B. D. McNicol, *Catal. Rev.*, *24* (2): 233 (1982)

5. R. C. Mackenzie (Ed.), *Differential Thermal Analysis*, Academic Press, London and New York (1970)

2

Theoretical Aspects

Although TPR is in no way restricted to studies of the hydrogen reductions of oxides, most of the work to date has been concerned with these systems. We therefore restrict the content of this chapter to refer to oxides and their reductions by hydrogen.

I. THERMODYNAMICS

The reaction between metal oxide (MO) and hydrogen to form metal (M) and water vapor can be represented by the general equation

$$MO(s) + H_2(g) \rightarrow M(s) + H_2O(g) \tag{1}$$

The standard free energy change for the reduction ΔG° is negative for a number of oxides (see Fig. 1), and thus for these oxides the reductions are thermodynamically feasible.

However, since

$$\Delta G = \Delta G^\circ + RT \log \frac{P_{H_2O}}{P_{H_2}}$$

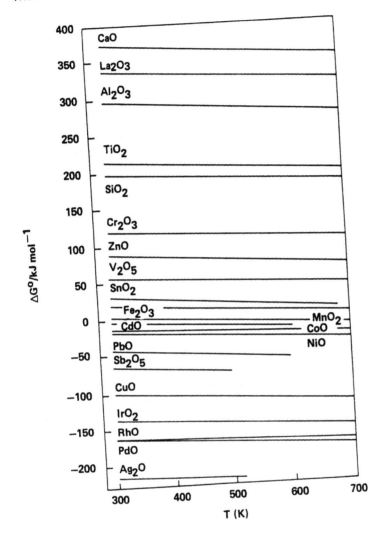

Fig. 1 Standard free energy change (ΔG^{o}) as a function of temperature for the process: Metal oxide + H_2 $\rightarrow$ metal + H_2O.

where P denotes partial pressure, it may still be possible for the reduction to proceed even when ΔG° is positive. The TPR experimental method is such that the water vapor is constantly swept from the reaction zone as it is formed. Thus, if P_{H_2O} is lowered sufficiently at elevated temperatures, it is possible that the term $RT \log (P_{H_2O}/P_{H_2})$ could be sufficiently negative to nullify a positive ΔG°. Thus, it has been possible to obtain TPR profiles for oxides of vanadium, tin, and chromium, which, at the reduction temperatures, have approximate positive ΔG° values (kJ/mol) of 45, 50, and 100, respectively (see Fig. 1).

II. KINETICS AND MECHANISMS OF REDUCTION

A. Bulk Oxides

Here we consider the process by which a sphere of metal oxide is reduced, directly to the metal, in a stream containing hydrogen [Eq. (1)] under isothermal conditions. It is common to observe the degree of reduction, α, as a function of time, t, for various temperatures and pressures of hydrogen. These data constitute the kinetics of the reduction which are interpreted in terms of the mechanism by which the reduction occurs. The data have been interpreted in terms of the nucleation model or the contracting sphere model. These are examined in turn below.

1. The Nucleation Model

When the oxide and hydrogen come into contact, the reaction starts and after some time, t_1, the first nuclei of the solid product form. Oxygen ions are removed from the lattice by reduction, and when the concentration of vacancies reaches a certain critical value they are annihilated by rearrangement of the lattice with the eventual formation of metal nuclei [1]. The reaction interface, which may be visualized as the interface between the nuclei of the metal and the metal oxide, begins to increase more and more rapidly by two processes: the growth of the nuclei already formed and the appearance of new ones. Oxygen ions may be removed by inward diffusion of hydrogen to the metal/metal oxide interface or outward diffusion of oxygen ions from the metal oxide to the metal/gas interface. At a certain stage of the reduction, the metal nuclei have grown at the surface of the oxide grains to such an extent that they begin to make contact with one another. From this moment a decrease of the reaction interface begins because of the overlapping of the metal nuclei and the steady consumption of the oxide grains. These processes continue until the oxide is completely reduced to the metal. The final stage, in which the reaction interface is shrinking as the metal layer grows, is equivalent to the contracting (or shrinking) sphere model, which may with some systems be operative from the onset of reduction (see below).

In some cases, for instance, with nickel oxide, the metal nuclei formed are thought to dissociate and activate the hydrogen so that the reduction is autocatalytic. Figure 2 illustrates this nucleation mechanism which results in the S-shaped α-against-t plot shown, and the maximum in the $d\alpha/dt$-against-α plot. In the figure, t_1 is the induction period (time) before the reaction starts. It should also be noted that a similar α-against-t curve is obtained for autocatalytic reductions.

A Mathematical Treatment of the Nucleation Model

We write [1,2]

$$V \propto (t - t_1)^p$$

where

  V = the volume of a nucleus

  p = the dimension of the nucleus

  t = time

  t_1 = the induction time

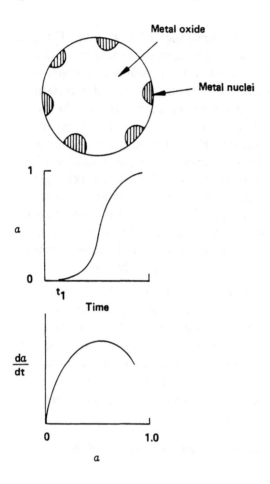

Fig. 2 Metal oxide reduction by a nucleation mechanism. α is the degree of reduction.

In addition

$$\frac{dG}{dt} \propto (t - t_1)^q$$

where

G is the number of nuclei

q is a number of value 0 or n, where n is an integer.

The total volume of product at time t (V(t)) is given by

$$V(t) = \int_{t_1}^{t} (dG/dt) \, V \, dt$$
$$= \text{const } (t - t_1)^{q + p + 1}$$

The const $= (C_1 C_2)/(q + p + 1)$, where C_1 and C_2 are proportionality constants. The degree of reduction, α, can be written in terms of the volume of product and the final volume

$$\alpha = V(t)/V_{final}$$

Therefore

$$\alpha = \frac{C_1 C_2}{V_{final} \, (p + q + 1)} (t - t_1)^{q + p + 1}$$

This equation gives the degree of reduction, α, as a function of time, t, for a nucleation process with nuclei of dimension p and a rate of nucleation proportional to $(\text{time})^q$. It represents the rising part of the α-vs.-time curve shown for the nucleation mechanism.

2. The Contracting Sphere Model

In many cases, the reaction interface decreases continuously from the beginning of the reaction when it has its maximum

value. This is usually interpreted in terms of a very rapid nucleation, which results in the total coverage of the oxide grain with
a thin layer of the product in the first instant of the reaction.
The reaction interface then decreases as the substrate grain is
consumed in the course of the reaction. The distinction between
the nucleation and contracting sphere models is somewhat artifi-
cial in that the contracting sphere model starts with very rapid
nucleation whereas the nucleation mechanisms finish by an essentially contracting sphere model. However, the distinction is
between a reaction interface (and hence rate) that is increasing
in the early stages of the reduction process and a reaction interface (and hence rate) that is contracting throughout the
reduction.

Figure 3 shows the contracting sphere model with its continuously decreasing metal/metal oxide interface which results in
the curved α-against-t plot shown, and the continuously decreasing rate throughout the reaction ($d\alpha/dt$ against t). Figure 3 represents a partially reduced oxide particle, comprising a sphere
of radius r_1 of metal oxide, surrounded by metal to a radius r_0.
The concentration of hydrogen is C_0 at r_0 and C_1 at r_1. C_{eq} is
the equilibrium concentration of hydrogen.

A Mathematical Treatment of the Contracting Sphere Model

The fractional thickness of the product layer, f, is defined

$$f = \frac{r_0 - r_1}{r_0} \text{ or } r_1 = r_0(1 - f)$$

If W = the weight of the product layer and W_0 = the weight of
the unreduced sphere, the degree of reduction

$$\alpha = W/W_0 = (1 - (1 - f)^3)$$

or

$$f = (1 - (1 - \alpha)^{1/3})$$

or

$$S/S_o = (1 - \alpha)^{2/3}$$

where

S = the metal/metal oxide interface area

S_o = the area of the unreacted oxide sphere

The model assumes that a reaction is taking place at the metal/metal oxide interface. The rate of this reaction may be controlled by either chemical or transport phenomena. It is possible to divide these major rate-limiting processes into many constituent parts; e.g., chemical control may be by diffusion from the gas stream to the particle or diffusion through the metal layer, which may or may not be porous. The many possible rate-limiting steps have been widely discussed [1,3-9].

We will consider here a very simple model where the reaction rate may be controlled either by the diffusion of H_2 to the metal/metal oxide interface or by the rate of the chemical reaction at this interface. This will serve to illustrate the type of analysis possible and the critical consideration that must be given to the mathematical statements of the various rate processes if comparison with experimental data is to have any significance.

The chemical rate at the interface may be written

$$\text{rate} = V_p = 4\pi r_1{}^2 k_p (C_1 - C_{eq}) \tag{2}$$

where k_p is a rate constant; i.e., the rate is proportional to the surface area of the metal/metal oxide interface, and first order with respect to hydrogen concentration.

The rate of diffusion of hydrogen through the metal layer may be written

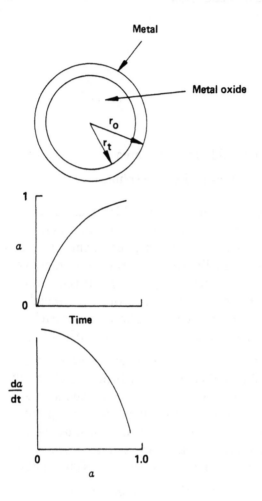

Fig. 3 Metal oxide reduction by a contracting sphere mechanism. α is the degree of reduction.

$$\text{Rate} = V_d = \frac{4\pi k_d (C_o - C_1) r_1 r_o}{(r_o - r_1)} \tag{3}$$

where k_d = a diffusion coefficient.

The steady-state approximation is now used so that

$$V_d = V_p$$

Then

$$V_p = \frac{4\pi r_1{}^2 k_p (C_o - C_{eq})}{1 + \dfrac{r_1}{r_o} \dfrac{k_p}{k_d} (r_o - r_1)} \tag{4}$$

since

$$W = \frac{4}{3} \pi d_o r_o{}^3 (1 - (1 - f)^3)$$

$$\text{rate } \frac{dW}{dt} = 4\pi d_o r_o{}^3 (1 - f)^2 \left(\frac{df}{dt} \right) \tag{5}$$

Equating (4) and (5), integrating, and using the substitutions $r_1 = r_o(1 - f)$ and $f = (1 - (1 - \alpha)^{1/3})$ yields

$$\frac{k_p}{r_o d_o} (C_o - C_{eq}) t = (1 - (1 - \alpha)^{1/3}) + \frac{r_o k_p}{k_d}$$

$$\left[\frac{1}{2} - \frac{\alpha}{3} - (1 - \alpha)^{2/3} \right]$$

This is the integrated rate equation which gives the degree of reduction, α as a function of time t.

If $k_p \ll k_d$ or $k_p/k_d \cong 0$, then

$$\frac{k_p}{r_o d_o} (C_o - C_{eq})t = (1 - (1 - \alpha)^{1/3})$$

A reduction process in which a plot of $(1 - (1 - \alpha)^{1/3})$ against t is a straight line is often interpreted as indicating chemical control at the metal/metal oxide interface. However, this conclusion should be treated with caution, since any process which results in a contracting sphere may give such a relationship, without any reference to the type of control, as a consequence of the geometry alone.

If $k_p \gg k_d$ or $k_d/k_p \cong 0$, then

$$\frac{k_d}{r_o d_o} (C_o - C_{eq})t = r_o \left[\frac{1}{2} - \frac{\alpha}{3} - \frac{(1 - \alpha)^{2/3}}{2} \right]$$

This is the equation for control of the overall reaction rate by diffusion of hydrogen through the product layer.

It is common practice to analyze experimental data in terms of either chemical or diffusion control, and if it is found that the results do not correspond exactly to either case, to infer that the reaction is under mixed control. However, the limitations of the model used must be carefully considered for such a conclusion to be valid; e.g., oxide ion diffusion is not considered in the model used here. Also, the chemical rate may not depend on the surface area, for example, where specific sites are active and the reaction may have a fractional, zero, or negative order with respect to hydrogen concentration. For the diffusion model, the effect of countercurrents of water vapor on the diffusion of hydrogen should be considered and the role of porosity of the metal layer may be important. The problem of increasing the complexity of the model is that many variables may be introduced that will inevitably lead to a fit of the model with the experimental α-against-t data without, however, any real significance.

B. Supported Oxides

Metal oxides supported on inert carriers such as γ-Al_2O_3 and SiO_2 may exhibit different reduction behaviors compared with the unsupported oxides. Reduction may be hindered or promoted, depending on the nature of the oxide/support interaction. This interaction may result in the formation of an identifiable metal aluminate or silicate [10]. Other authors [11,12] have expressed the interaction between metal oxide and alumina as one that results in a substantial fraction of the supported metal oxide having an oxidation state of the metal above its normal value.

Supported metal oxides may be homogeneously distributed across the surface of the support or exist as islands of oxide separated by uncovered support. Islands of metal oxides may be expected to reduce in a similar manner to unsupported oxide, in which case the support may act purely as a dispersing agent and promote the reduction. Under these conditions the reduction kinetics observed for the supported oxide are of the same general form as those found for the bulk oxide. For example, both bulk [13] and supported [14] CuO reduce with the sigmoidal α-against-t plot characteristic of a nucleation of autocatalytic reduction (Fig. 2), and supported CuO is found to reduce more easily.

Bulk NiO [7,15,16] can show both nucleation and contracting-sphere-type reduction isotherms, while supported NiO, which is more difficult to reduce than the bulk oxide, shows a reduction isotherm that is characterized by constantly decreasing rates of reaction [17,18] as shown for the contracting sphere model (see above). The concept of a contracting sphere would seem to be inappropriate in the case of a homogeneously distributed supported oxide, so that the reduction may involve instantaneous nucleation across the whole area of the supported oxide.

Metal atoms and crystallites are known to be mobile on the surface of supported metal oxides [19] so that under

appropriate conditions the reduction of a homogeneously supported metal oxide may proceed with the reduction of individual
metal ions, or groups of metal ions, followed by surface diffusion
to form metal crystallites which in turn may diffuse and com-
bine to form particles of the reduced phase. Such a reduced
phase may have an autocatalytic effect on the reduction.

The nucleation processes may, however, be hindered by
metal/support interactions reducing the mobility of metal atoms.
Furthermore, any autocatalytic effect of the reduced phase may
be hindered if activated hydrogen species are not mobile across
the support surface at the reduction temperature. Hydrogen
atoms are known to spill over from supported Pt onto an Al_2O_3
support only above 600 K [20].

C. Zeolites

Zeolites containing reduced metal, whether atomically dispersed
or as small clusters within the pore structure, are of great value
in the petrochemical industry as hydrocracking and isomerization
catalysts [21,22]. The activity of these catalysts is critically dependent on both the degree of dispersion and the location of
the metal, which are in turn dependent on the method of reduction and pretreatment. Thus, an understanding of the kinetics
and mechanism of reduction is particularly valuable in the preparation of these catalysts.

The reduction of a metal ion in a zeolite is generally achieved
by using hydrogen; the overall reaction stoichiometry may be
written [23]

$$M^{n+} + \frac{n}{2} H_2 \rightarrow M^\circ + nH^+ \tag{6}$$

The protons react with the zeolite lattice to produce the hydroxyl groups whose presence has been established by IR spectroscopy [24]:

$$nZO^- + nH^+ \rightarrow nZOH \tag{7}$$

The reduction of transition metal ions in zeolites has been reviewed by Uytterhoeven [23], who observed a correlation between the reducibility and standard electrode potential. His data are reproduced in Table 1 along with more recent TPR results (see Chapter 4). Care should be taken in comparing the results of different authors since it is known that the TPR peak positions are sensitive to experimental conditions (see Section IV). However, the sequence obtained by each author is in general agreement with the prediction.

Detailed kinetic studies have been limited to a few systems. Beyer and colleagues [24] observed two kinetic regimes in the reduction of silver ions in Y zeolite

(i) $T < 430$ K

$$\frac{dc}{dt} = K' P(C_0 - C)/C$$

$E = 40$ kJ mol^{-1}

(ii) $T > 430$ K

$$\frac{dc}{dt} = K_1 (C_1 - C)$$

$E = 97.6$ kJ mol^{-1}

where

C_0 = initial concentration of Ag^+

C = concentration of Ag^+ reduced at time t

P = hydrogen pressure

Beyer et al. [24] suggest a mechanism that involves the activation of hydrogen by impurity paramagnetic centers (presumably Fe^{3+}) and diffusion of the metal ion. Thus

$$S + H_2 \rightarrow SH^- + H^+ \tag{8}$$

$$SH + Ag^+ \rightarrow SAgH \tag{9}$$

$$SAgH + Ag^+ \rightarrow Ag_2 + S + H^+ \tag{10}$$

where S is an impurity paramagnetic center.

Table 1   Reducibility of Transition Metal Ions in Zeolites Y and X

Reduction process	Standard electrode potential (V)	Temperature (K) of first major reduction maximum (minor peaks in parentheses)	
		Zeolite Y	Zeolite X
Pd^{2+}/Pd^{o}	+0.83	290	250
Ag^{+}/Ag^{o}	+0.80	393	
Cu^{2+}/Cu^{+}	+0.16	440; 520	550; 540
Cu^{+}/Cu^{o}	+0.52	700; >773	730; 773
Ni^{2+}/Ni^{o}	−0.23	773; (600) 950	(600) 900; (570) 830
Co^{2+}/Co^{o}	−0.28	a	a
Cd^{2+}/Cd^{o}	−0.40	a	
Fe^{2+}/Fe^{o}	−0.41	a	700
Cr^{3+}/Cr^{2+}	−0.41		650
Cr^{3+}/Cr^{o}	−0.74		
Zn^{2+}/Zn^{o}	−0.76	a	a
Mn^{2+}/Mn^{o}	−1.03	a	a

[a]No reduction observed.

At low temperatures, reduction is limited to two-thirds of the silver available in the supercage. Beyer et al. [24] propose that the reduced silver atoms become associated with Ag^+ to form Ag_3^+ species which are resistant to further reduction at these temperatures. The species were found to be stable to sintering at 623 K and could be fully reoxidized to their original ionic state.

At higher temperatures complete reduction is achieved. It is proposed that the migration of Ag^+ from S_1 sites to activated hydrogen sites becomes rate limiting and that migration of silver clusters to the external surface follows to produce crystallites ~ 20 nm diameter.

From a study of the reduction of silver mordenite, Beyer and Jacobs [26] concluded that a similar mechanism prevailed and that the dominant cluster produced in the main channels corresponded to Ag_5^+.

Jacobs and co-workers [27] have studied the reduction of NiY and shown that after incomplete dehydration (in vacuo at ambient for 2 h followed by slow heating to 373 K) the degree of reduction increased with temperature up to 100% at 873 K. After a more thorough dehydration [24], it was found that reduction was restricted to about 30% at 873 K, suggesting that water enhances the reduction process. The kinetics could be explained in an analogous way to that proposed by Beyer et al. for AgY, that is:

1. Activation of hydrogen at some surface site

2. Diffusion of Ni^{2+} toward the active site

3. Regeneration of the site as the rate-determining step

4. The formation of charged clusters, estimated to be Ni_3^{2+} or Ni_4^{2+}

Briend-Faure et al. [28] studied the reduction of Ni^{2+} in both X and Y zeolites. In NiY they observed three distinct kinetic regimes dependent on both the initial level of exchange and the degree of reduction, and they correlated these reduction regimes to the initial distribution of Ni^{2+} ions. The results are summarized in Table 2.

All the foregoing work on Ni^{2+} zeolite systems has found Ni^o as the only reduction product. However, Garbowski and colleagues [29,30], using EPR and electron spectroscopy, have observed Ni^+. TPR at relatively high hydrogen pressure [23] was unable to identify a two-step mechanism; hence, it must be assumed that the concentration of Ni^+ is low and that further reduction to Ni^o is rapid.

In contrast with Ni^{2+} the reduction of Cu^{2+} to Cu^+ can be achieved almost stoichiometrically [31,25]. Two separate $Cu^{2+} \rightarrow Cu^+$ reductions are observed. Jacobs et al. [31] propose a mechanism

$$
\begin{array}{cc}
\overset{\displaystyle H\text{-----}H}{\underset{\displaystyle Cu^{2+} + O_1{}^- + H_2 \rightleftharpoons Cu^{2+}}{\vdots\qquad\qquad\vdots}} & \overset{\displaystyle }{\underset{\displaystyle O_1{}^- \rightleftharpoons (Cu^{2+}H^-) + (O_1{}^-H^+)}{}}
\end{array}
$$

$$\text{(11)}$$

$$(O_1{}^-H^+) \rightleftharpoons O_1H \tag{12}$$

$$(Cu^{2+}H^-) + Cu^{2+} \rightarrow 2CU^+ + H^+ \tag{13}$$

where O_1 is a lattice oxygen atom and diffusion of Cu^{2+} is rate limiting. The two processes observed are believed to be diffusion and reduction of supercage ions ($E = 70$ kJ mol^{-1}) and migration of sodalite ions into the supercages followed by reduction ($E = 111$ kJ mol^{-1}). A small amount of Cu^{2+} remaining unreduced is assumed to occupy hexagonal prism sites.

Gentry et al. [25] have used TPR to identify the two separate processes for the reduction of $Cu^{2+} \rightarrow Cu^+$ in CuY, with

Table 2   Kinetic Parameters for Reduction of Ni^{2+} Ions in Various Lattice Sites in X and Y Zeolites

Activation energy, E (kJ mol^{-1})	Degree of reduction (α)		Ni^{2+} Sites[a]
	20% Exchanged	13% Exchanged	
104	<0.15		II
134	$0.15 < \alpha < 0.35$	<0.13	I$'$
167	>0.35	>0.13	I

[a] II indicates the supercage, I$'$ the sodalite cage, and I the hexagonal prism.
Source: Ref. 28.

activation energies of 64 and 84 kJ mol^{-1}, respectively, in broad agreement with Jacobs et al. Furthermore, two identical processes were identified in CuX. Gentry and colleagues suggest that an alternative mechanism may be involved in which Cu^{2+} migration from I$'$ to II is not a rate-limiting step. Support for this is found in the work of Leach et al. [32,33], where adsorption of H_2 was found not to induce ion migration.

Possible reaction schemes may be

$$2ZO^- + Cu^{2+} + H_2 \quad \rightarrow 2ZO^- (Cu^{2+} - H_2) \tag{14}$$

$$2ZO^- (Cu^{2+} - H_2) \quad \rightarrow ZOH + ZO^-Cu^+ + H^\circ \tag{15}$$

$$\rightarrow ZO^-H_2^+ + ZO^-Cu^+ \tag{16}$$

$$2ZO^- + Cu^{2+} + H^\circ \quad \rightarrow ZOH + ZO^-Cu^+ \tag{17}$$

$$2ZO^- + Cu^{2+} + ZO^-H_2^+ \rightarrow 2ZOH + ZO^-Cu^+ \tag{18}$$

where the formation of the $(Cu^{2+} - H_2)$ intermediate or migration of either hydrogen species (H_2^+ or H^o) may be the rate-controlling process. However, Jacobs et al. [31] report that no kinetic isotope effect was observed when deuterium was used as the reducing agent.

The stability of Cu^+ reported above is not expected from the standard electrode potential (Table 1). Indeed, Texter and co-workers [34] observed that Cu^+ introduced by ion exchange with CuI in liquid ammonia reduced more readily than the Cu^{2+} species. It is therefore interesting to speculate on the stability of the Cu^+ species formed on partial reduction of Cu^{2+}.

Since the reduction processes involve the formation of hydroxyl groups, it is suggested that the stability arises from the inhibiting influence of acid groups adjacent to the Cu^+ species. Thus further reduction of Cu^+ produced from initial Cu^{2+} is hindered by the close proximity of the product hydroxyl groups [see Eqs. (12), (15), etc.], while the low acidity of the ion-exchanged Cu^+ species offers no such inhibition. Furthermore, the lower Si/Al ratio in X zeolite allows a greater density of hydroxyl groups than Y zeolite, which may explain the formation of some Cu^o in CuX under the mild conditions used by Gentry et al. [25]. This "self-inhibiting" effect may also have some influence on the observed cluster sizes in AgY and NiY.

Some general conclusions may be drawn from the above brief review.

1. The ease of reduction of different ions follows the standard electrode potential series, although unexpected oxidation states may be stabilized by the zeolite lattice.

2. The ease of reduction is dependent on the location of the ion:

supercage $>$ sodalite cage $>$ hexagonal prism

(II)   (I′)   (I)

3. The rate-determining step is generally associated with the migration of ions to reduction sites.

4. Reduction is frequently limited to less than 100% with the formation of electron-deficient clusters.

D. Bimetallics

In this section we consider the more important aspects of the promoting influence of foreign metal species on the reduction of bulk metal oxides. This includes the autocatalytic effect of the reduced phase on the reduction of its oxide. These subjects have been widely studied [1,17,35-41]. For convenience, the term *dopant* is used here to signify the foreign metal species added to the bulk oxide.

A dopant may promote reduction either by lowering the activation energy of the rate-determining step or by increasing "preexponential" terms. The activation energy for the reduction of CuO was not significantly altered by the addition of Pd or Pt [40], although a strong promoting effect was observed. Similarly [37], the activation energies of reduction of Fe_2O_3, UO_3, NiO, and MnO_2 were not significantly different when Pt was used to promote the reduction.

Accordingly, the promoting effect of metal dopants would seem to arise from increases in the preexponential terms. These can be produced in two ways: either by directly increasing the number of nucleation or potential nucleation sites or by providing a high concentration of active hydrogen which is efficient at causing nucleation. These two methods amount, directly or indirectly, to promotion of the nucleation stages of reduction.

For a doped metal oxide it has been suggested [17] that nucleation can occur at four different types of site:

i. At undisturbed parts of the surface. This is natural, or spontaneous, nucleation and would correspond to a similar mechanism on untreated samples.

ii. At disorganized parts of the surface, i.e., portions that have been perturbed by the incorporating process but where no dopant is present.

iii. At portions where the parent metal ion has been replaced by the dopant ion.

iv. At portions where the dopant, representing a distinct new phase, is in contact with the metal oxide.

For iii and iv, nucleation occurs in contact with the dopant ion, and the ease of reducibility of the dopant ion, compared to the parent ion, is also of major importance. If the dopant ion reduces first, it may produce a high concentration of active hydrogen on the surface which may promote the reduction.

Reduction is called autocatalytic when reduced metal formed in the early stages of reduction catalyzes the reduction by H_2 activation. This produces the S-shaped (or sigmoidal) α-against-t reduction isotherm familiar for nucleation and autocatalytic reduction (e.g., the reduction of CuO and NiO). In the presence of hydrogen-activating dopants, these oxides still reduce with an S-shaped reduction isotherm [35]. It would appear in these examples that the role of the dopant is that of hydrogen activator; i.e., the nucleation steps are indirectly promoted. Here the distinction between mechanisms iii and iv given above may become critical [40]. If the dopant is incorporated into the lattice, it may activate hydrogen which can diffuse by intraparticular spillover. This process may be expected to be very efficient. However, if the dopant exists as a distinct phase, then the hydrogen must diffuse by the less efficient interparticular spillover (or jump-over) mechanism. This can greatly affect the promoting influence of a metal on the reduction of an oxide [40]. First row transition metals (Cr, Mn, Fe, Co, Ni) have also been found to promote the reduction of CuO [40]. In such cases the dopant ion is not thought to reduce before the parent ion, and the promoting effect is thought to arise due to increased nucleation at sites such as ii and iii above.

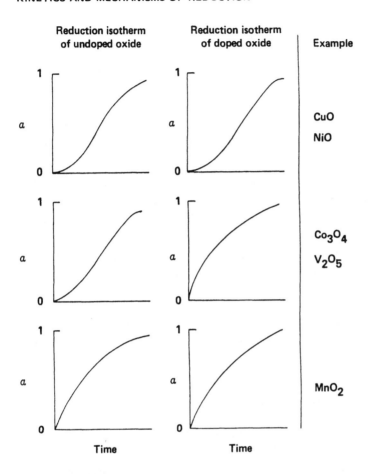

Fig. 4 Kinetics of reduction of doped and undoped oxides. α is the degree of reduction.

In contrast to this, other oxides (Co_3O_4 and V_2O_5), which also produce S-shaped reduction isotherms characteristic of a nucleation mechanism when undoped, produce reduction isotherms characteristic of a contracting sphere model when in the presence of hydrogen-activating catalysts, e.g., Pt [37]. It would seem that in these cases it is the role of the dopant as a nucleation catalyst which is important; that is, the nucleation steps are directly promoted.

Indeed, oxides that reduce by a contracting sphere model when undoped cannot have their reductions promoted [17]; for instance, the reduction of MnO_2 to Mn metal, which is characterized by a continuously decreasing rate, is not promoted by the presence of many substances (Ti, Cr, W, Re, Fe, Co, Ni, Ru, Rh, Pd, Os, Ir, Pt, Cu, Ag). It would appear that the undoped oxide reduces by a spontaneous nucleation over the entire surface immediately on contact with hydrogen, and the presence of a dopant metal is unable to promote the reduction by either nucleation or hydrogen activation. These results are summarized in Fig. 4.

III. DERIVATION OF KINETIC PARAMETERS

The aim of the analysis of TPR data has been to derive kinetic parameters relating to the reduction process. In general, analyses have used methods originally developed for other temperature-programmed techniques (42,43].

A. Derivation from TPR Profiles Obtained at Different Heating Rates

Banerjee et al. [44,45] have used the method of Coats and Redfern [46] originally developed for thermogravimetric data to derive kinetic parameters from a study of the changes in the solid phase in excess reducing gas.

For the reaction

$$\text{Gas + solid} \rightarrow \text{products} \tag{19}$$

The rate of reaction of the gas [G] or solid [S] at constant temperature may be expressed as

$$\text{Rate} = \frac{-d[G]}{dt} \text{ or } \frac{-d[S]}{dt} = k[G]^p[S]^q \tag{20}$$

where the rate constant k is given by the Arrhenius equation

$$k = Ae^{-E/RT} \tag{21}$$

In temperature-programmed reduction, temperature is also a function of time. Thus

$$\beta = \frac{dT}{dt} \tag{22}$$

where β is the linear heating rate, so that

$$\frac{-d[G]}{dt} = \frac{-\beta d[G]}{dT} \tag{23}$$

or

$$\frac{-d[S]}{dt} = \frac{-\beta d[S]}{dT} \tag{23a}$$

In excess reducing gas the reaction is independent of gas concentration, and Eq. (20) is rewritten

$$\frac{d\alpha}{dt} = k(1 - \alpha)^q \tag{24}$$

where α is the fraction reduced at time t. Combining Eqs. (21), (22), and (24), rearranging, and integrating gives

$$\int_0^\alpha \frac{d\alpha}{(1 - \alpha)^q} = \frac{A}{B} \int_0^T e^{-E/RT} \, dT \tag{25}$$

using an approximation [47] for the right-hand integral gives, for all values of q except q = 1,

$$\frac{1 - (1 - \alpha)^{1-q}}{1 - q} = \frac{ART^2}{\beta E} \left[1 - \frac{2RT}{E}\right] e^{-E/RT} \tag{26}$$

Taking logs gives

$$\log \left[\frac{1 - (1 - \alpha)^{1-q}}{T^2(1 - q)}\right] = \log \frac{AR}{\beta E} \left[1 - \frac{2RT}{E}\right] - \frac{E}{2.3RT} \tag{27}$$

When q = 1, Eq. (25) after taking logs becomes

$$\log \left[-\log \frac{(1 - \alpha)}{T^2}\right] = \log \frac{AR}{\beta E} \left[1 - \frac{2RT}{E}\right] - \frac{E}{2.3RT} \tag{28}$$

It can be shown that for most values of E over the temperature range of these experiments the expression $\log (AR/\beta E) [1 - (2RT/E)]$ is essentially constant. Thus plotting the left-hand side of Eq. (27) or (28) against 1/T for the correct value of q yields a straight line of slope $-E/2.3R$. This method of analysis is not, however, generally useful as an aid to the interpretation of TPR data since most experimental TPR systems monitor the rate of change of gas concentration as a function of temperature, producing reduction profiles with peaks corresponding to maxima for the rate processes.

Thus, a more generally useful analysis by Gentry et al. [25] started with a statement of the reaction rate in terms of the rate of consumption of gas for a reaction taking place between a solid held in a tubular reactor, at constant temperature, and a gas flowing through it. Provided that the flow can be described as plug flow (i.e., radial and longitudinal mixing are absent) the rate of consumption of gas, at low conversions, is given [48] by

Rate = fx

where f is the feed rate of the gas and x is the fractional conversion of the gas.

Now

$$f = F[G]_i$$

where F is the flow rate and $[G]_i$ is the concentration of gas G at the reactor inlet
and

$$x = \frac{[G]_i - [G]}{[G]_i} = \frac{\Delta[G]}{[G]_i}$$

where $[G]$ is the concentration of gas at the reactor outlet. Thus, the rate can be measured as

$$\text{Rate} = F\Delta[G] \tag{29}$$

By using a recirculation reactor (see Chapter 3, Section III), Jacobs and colleagues [49] obtained reaction rates directly by monitoring the reaction pressure as a function of time.

Differentiating Eq. (20) with respect to temperature gives

$$\frac{d(\text{rate})}{dT} = A \exp(-E/RT) \left[[G]^p[S]^q \frac{E}{RT^2} \right. \tag{30}$$
$$\left. + q[G]^p[S]^{q-1} \frac{d[S]}{dT} + p[G]^{p-1}[S]^q \frac{d[G]}{dT} \right]$$

at the maximum rate

$$\frac{d(\text{rate})}{dT} = 0 \tag{31}$$

and from Eq. (29)

$$\frac{d[G]}{dT} = 0 \qquad (32)$$

Denoting parameters relating to the maximum rate by the subscript m and simplifying, Eq. (30) becomes

$$\frac{E}{RT_m^2} + \frac{q}{[S]_m} \frac{d[S]_m}{dT} = 0 \qquad (33)$$

Combining Eqs. (20), (21), (23a), and (33) gives

$$\frac{E}{RT_m^2} = \frac{A[G]_m^P \, q[S]_m^{q-1}}{\beta} \exp(-E/RT_m)$$

Taking logs and rearranging gives

$$\ln \frac{T_m^2[G]_m^P}{\beta} + \ln q[S]_m^{(q-1)} = \frac{E}{RT_m} + \ln \frac{E}{RA} \qquad (34)$$

Thus, if the values of p and q are known, a plot of the left-hand side of Eq. (35) against $1/T_m$ yields a straight line of slope E/R and intercept ln E/RA. Both E and A can be obtained in this way.

The reaction order with respect to reducing gas and solid may, in theory, be derived by substitution of values of p and q into Eq. (34). However, orders with respect to the solid phase are frequently fractional and expressions such as Eq. (34) may be insensitive to the value of q. Consequently, exact derivation of a value of q is not possible. If it is assumed that q = 1, then

$$\ln \frac{T_m^2[G]_m^P}{\beta} = \frac{E}{RT_m} + \ln \frac{E}{RA} \qquad (35)$$

The value of E derived from this expression is dependent on the chosen value of p. Thus, for the reduction of CuO [40] it was found for a zero order reaction in hydrogen, E was a function of hydrogen concentration, whereas for p = 1 a unique value of E was obtained for all concentrations used.

In a similar analysis, Monti and Baiker [50] considered that the use of a mean hydrogen concentration was more appropriate to a reaction occurring in a differential reactor. For the case of first order kinetics (i.e., p = q = 1) these authors derived an expression

$$\ln \frac{T_m^2 \overline{C}_m}{\beta} = \frac{E}{RT_m} + \ln \frac{E}{RA} \tag{36}$$

where $\overline{C}_m$ is the mean hydrogen concentration at the temperature of the maximum reduction rate. At low conversions where $[G]_i \approx [G]$ the two equations are identical.

B. Derivation by Analysis of a Single TPR Profile

1. Curve Fitting

By using a similar set of equations as used above [Eqs. (20), (21), (22), and (29)] for the case where reaction orders are unity (p = q = 1) Monti and Baiker [50] derived the expression:

$$\frac{d[G]}{dT} = \frac{d[S]}{dT} = \frac{2[G]_i F}{\beta} \left[1 + \frac{2F}{SA\exp(-E/RT)} \right]^{-1} \tag{37}$$

Thus, integration of Eq. (37) gives the relation between the hydrogen consumption rate and the sample temperature.

The sample temperature and the difference in inlet and outlet hydrogen concentration were normalized by use of the equations

$$T_n = T/T_m \tag{38}$$

$$[G]_n = ([G]_i - [G])/([G]_i - [G]_m) \tag{39}$$

and

$$E_m = E/RT_m \tag{40}$$

By numerical integration of equation (37) and standardization with Eqs. (38) and (39), Monti and Baiker evaluated normalized TPR profiles of $[G]_n$ as a function of T_n. Illustrative curves are reproduced in Figs. 5a and 5b. From these curves Monti and Baiker established that the only parameter that determines the peak shape (for this simple system) is E_m and that the asymmetric peaks become sharper as the value of E_m increases.

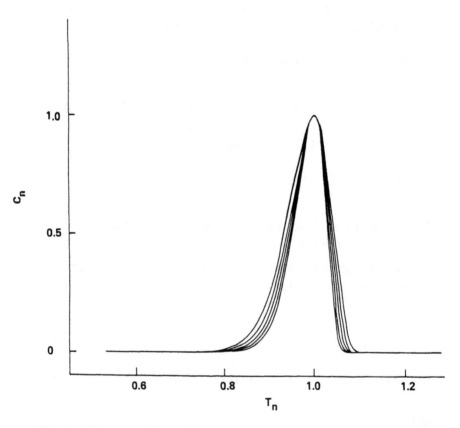

Fig. 5a Dimensionless TPR peak shape analysis for different values of E_m. E_m = 22.0; 24.4; 26.8; 29.2; 31.6. (From Ref. 50.)

The estimation of values for A and E were carried out by using a computer program for the regression analysis of single- and multiresponse algebraic and ordinary differential equation models. The computer program numerically integrated Eq. (37) and estimated values of A and E to give the best fit of the theoretical curve to the measured TPR peak. Results for the hydrogen reduction of nickel oxide in good agreement with published values were obtained. This method has the advantage that

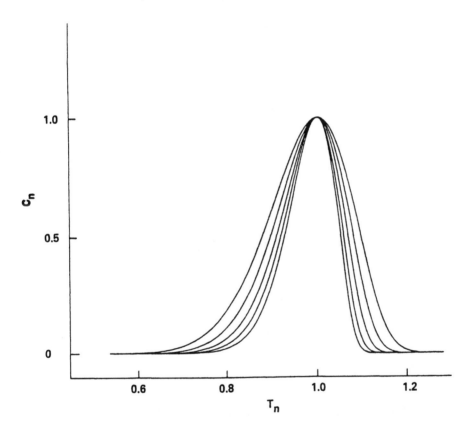

Fig. 5b Dimensionless TPR peak shape analysis for different values of E_m. The peaks become shaprer with increasing E_m. E_m = 10.4; 12.7; 15.0; 17.3; 19.7. (From Ref. 50.)

only one reduction profile is needed for an estimation of kinetic parameters. The authors point out that although it has been applied to a system with simple kinetics (first order with respect to gas and solid phase), it is possible, by applying numerical methods, to develop similar data analysis for complex kinetic rate expressions.

Simulations of TPR curves incorporating particle size distributions has been attempted by Tonge [16]. The results of this work are described in Section IV below.

2. Other Analyses of Single TPR Profiles

In principle, methods of peak analysis developed for other temperature-programmed techniques can be applied to TPR profiles, although few applications have been reported. Such methods [42] include peak width analysis for the determination of activation energy and shape index and slowness parameter analysis for evaluation of reaction orders.

Tonge [16] has employed the peak width analysis method using a previously established isothermal rate law in a study of the simulation of the TPR of nickel oxide.

IV. EFFECTS OF VARIATION OF EXPERIMENTAL PARAMETERS

TPR measurements reported in the literature have been taken over a wide range of experimental conditions. It is vital if a meaningful comparison is to be made between different studies to be clear on the sensitivity of the results to the experimental conditions used.

Most of the data presented below are concerned with the sensitivity of the temperature T_m to changes in the operating variables.

A. Influence of Heating Rate and Initial Temperature

The variation of T_m with heating rate β is most commonly used to obtain activation energies for the reduction process (Section

III.A). When the change in T_m is due to kinetic parameters as described in Eq. (35) above, other parameters remaining constant, the sensitivity of T_m to changes in heating rate is established.

Thus, in a systematic study of the hydrogen reduction of nickel oxide (a simple, one-step reduction, first order in hydrogen and nickel oxide), Monti and Baiker [50] found that an increase in heating rate from 0.09 to 0.31 Ks^{-1} produced an increase in T_m of 33 K. As can be seen from their results presented in Fig. 6, calculated values using estimated values of E and A were in good agreement with the experimental results. As expected, the gradient of the curve decreases with increasing values of β.

The influence of changes in the initial temperature, T_0, was considered in analysis by Tonge [16] (see Section IV.F below). It was calculated that a difference of 10K at a T_0 of 300 K shifted a typical peak in the 700-800 K region by only 2 K. Thus, normal changes in room temperature will not introduce significant errors.

An illustration of how changes in the chemistry of a reduction can complicate the simple kinetic effects comes from a study of the TPR of the hydrogen reduction of V_2O_5. Bosch and co-workers [51] found that for this reduction the heating rate influenced not only the peak positions but also the form of the TPR profiles (see Fig. 7). A heating rate of less than 5 K min^{-1} resulted in poor resolution of the first two peaks, and an extra peak was observed at heating rates above 10 K min^{-1}. At 16 K min^{-1} the change in the TPR profile was even more pronounced. These results were explained by a consideration of complexity of the reduction (see Chapter 4) and the influence of water production on the reduction process.

B. Influence of Reducing Gas Concentration

Figure 8 shows the results obtained by Monti and Baiker [50] for the hydrogen reduction of nickel oxide. An increase in hydrogen concentration from 3%v (1.23 μmol cm^{-3}) to 15%v

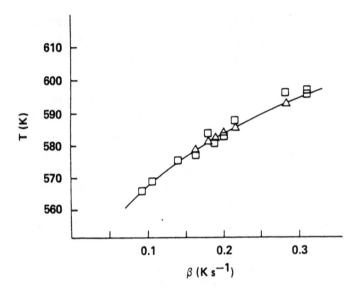

Fig. 6 Parametric sensitivity of the temperature of the maximum reduction rate for the hydrogen reduction of nickel oxide.
Influence of heating rate, β. Standard conditions: hydrogen
concentration, 2.46 μmol cm^{-3}; mass of sample, 405 μmol; flow
rate, 1.25 cm^3 (NTP) s^{-1}. The NiO mean particle size was
13.5 μm. Only particles of less than 100 μm were used.
□ Experimental values; △ Values calculated from the integration
of Eq. (37). (From Ref. 50.)

(6.15 μmol cm^{-3}) produced a decrease in T_m from 604 to 563
K in line with expections from the kinetic analysis given above
(Section III.A.) for this simple one-step reduction. The same
authors used the variation of T_m with hydrogen concentration
to derive values for the activation energy of the reduction using
Eq. (36).

C. Influence of Gas Flow Rate

The influence of gas flow rate has been investigated by Gentry
et al. [25] for the hydrogen reduction of Cu-exchanged zeolite.

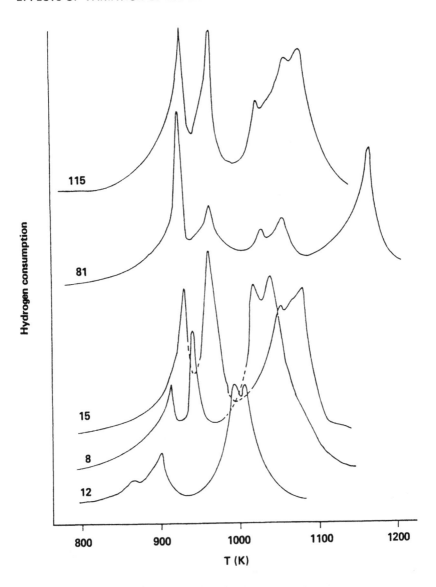

Fig. 7 Influence of heating rate on TPR-spectra for hydrogen reduction of V_2O_5. Run 12: 14.8 mg sample weight, 4.6 K min^{-1}; run 8: 16.2 mg, 5.6 K min^{-1}; run 15: 14.9 mg, 9.8 K min^{-1}; run 81: 10.9 mg, 16.4 K min^{-1}; run 115: 4.3 mg, 16.1 K min^{-1}. Volume flow rate, 10 cc min^{-1}, hydrogen concentration 6% H_2. (From Ref. 51.)

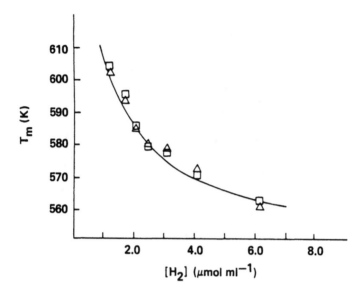

Fig. 8 Parametric sensitivity of the temperature of the maximum reduction rate. Influence of initial hydrogen concentration $[H_2]$. Standard conditions: mass of sample, 405 μmol; flow rate, 1.25 cm^3 (NTP) s^{-1}; β, 0.2 K s^{-1}. The NiO mean particle size was 13.5 μm; only particles of less than 100 μm were used. □ Experimental values; △ calculated values from the integration of Eq. (37). (From Ref. 50.)

They found that an increase in flow rate of 4%vH_2 from 10 to 20 mL min^{-1} at constant total pressure lowered the value of T_m by 15-30 K.

Flow reactor theory [48] shows that an increase in flow rate for a reactant consumed by a first-order process results in a lowering of the degree of conversion and hence an increase of reactant concentration in the reactor. Experimental results showed that the above increase in flow rate increased $[H_2]_m$ by $\approx 1\%$. Equation (35) predicts that such a change in $[H]_m$ will produce a decrease in T_m of ≈ 15 K, in broad agreement with the observed result.

For the hydrogen reduction of nickel oxide, another first-order reduction in hydrogen, Monti and Baiker [50] also found that an increase in flow rate lowered the value of T_m. Their results are shown in Fig. 9. Increasing the flow rate from 0.7 to 1.7 cm^3 (NTP) s^{-1} has only a small effect on T_m. The difference between the experimental and calculated results at lower flow rates was attributed to inordinate time lags for the temperature measurement and an increase in dispersion effects (see below). The authors argue that with proper selection of operating parameters such discrepancies do not arise (see Section IV.E below).

Bosch et al. [51] demonstrated that doubling the flow rate from 10 to 20 mL min^{-1} had no effect on the peak positions during a TPR study of the hydrogen reduction of V_2O_5, from which they concluded that the order of the reduction rate in hydrogen is low or even zero.

1. Dispersion Effects on Reduction Profiles

As indicated above, errors can arise, particularly at low flow rates, from the spreading of gas "peaks" between the reactor and detector. This effect was investigated by Monti and Baiker [50] in an apparatus in which the reactor was built according to a design by Cvetanovic and Amenomiya [52] in which volume of the tubing between reactor and detector was kept to a minimum. They constructed a special system that allowed the injection of a small quantity of helium tracer gas directly into the reactor in a predetermined manner. Using a tanks-in-series model the measured residence time distribution, $f(t)_1$, was described, and from measured TPR profiles the function of the test peak was described by a gaussian distribution $g_{in}(t)$. The signal at the detector, $g_{out}(t)$, was calculated from the convolution integral of $f(t)$ with $g_{in}(t)$. From a comparison of the functions g_{in} and g_{out} (presented in Fig. 10) Monti and Baiker were able to establish conditions under which the peak shapes were not perturbed by dispersion effects in their experiments.

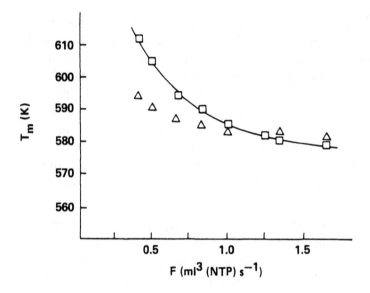

Fig. 9 Parametric sensitivity of the temperature of the maximum reduction rate. Influence of total flow rate. Standard conditions: mass of sample, 405 μmol; hydrogen concentration, 2.46 μmol cm^3; β, 0.2 K s^{-1}. The NiO mean particle size was 13.5 μm, only particles of less than 100 μm were used. □ Experimental values; △ calculated values from the integration of Eq. (37). (From Ref. 50.)

D. Influence of the Mass of Solid

Theory predicts that T_m should be independent of the mass of the solid and indeed in their study of the hydrogen reduction of nickel oxide Monti and Baiker [50] found only a minimal change in T_m for an increase in sample mass from 200 to 500 μmol (see Fig. 11).

In the study of the reduction of V_2O_5, Bosch et al. [51] found that changing the sample size from 5 to 43 mg had no effect on the first two peaks of the TPR spectrum but that for sample sizes of more than 20 mg the third and fourth peaks

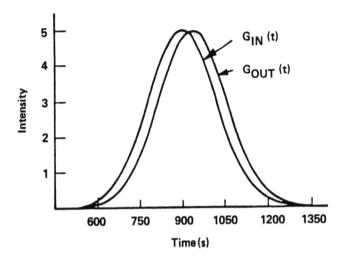

Fig. 10 Influence of the dispersion caused by the tubing between the reactor and the detector on the peak shape of a test
peak. (From Ref. 50.)

moved to higher temperatures. The effect was explained by the
effects of enhanced water vapor production with high sample
masses.

In a systematic study of the effects of sample mass Gentry
et al. [25] obtained TPR profiles at $\beta = 9.1$ K min^{-1} using
$4\%vH_2$ for masses of (Cu, Na)-X-50 between 50 and 400 mg.

The main effect of changing the mass was that the resolution of two separate reduction processes obtained with a 50-mg
sample was completely lost when the mass of zeolite was increased to 400 mg. In addition, the value of T_m for the com-
posite reduction peak was higher than those of the two separate
reduction peaks.

In order to determine whether these effects were artifacts
of the experimental system or due to chemical modification during pretreatment of large zeolite masses, so-called deep-bed calcination, a further series of experiments was performed on

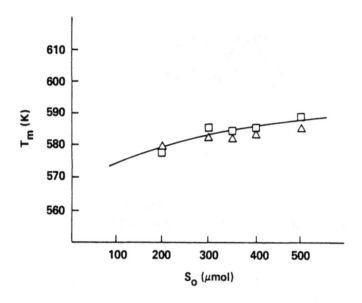

Fig. 11 Parametric sensitivity of the temperature of the maximum reduction rate. Influence of the mass of solid, So. Standard conditions: hydrogen concentration, 2.46 μmol cm^3; β, 0.2 K s^{-1}; flow rate, 1.25 cm^3 (NTP) s^{-1}. The NiO mean particle size was 13.5 μm; only particles of less than 100 μm were used. $\square$ Experimental values; $\triangle$ calculated values from the integration of Eq. (37). (From Ref. 50.)

various masses of (Cu,Na)-Y-68. This zeolite was chosen because it was shown to have reversible redox behavior. A similar loss of resolution was observed as the mass was increased from 50 to 400 mg. After reduction of a 1-g sample a 50 mg batch was removed, reoxidized at 773 K, and re-reduced. The resulting TPR profile was identical with that for the reduction of a fresh 50-mg sample. Thus, the loss of resolution could not be attributed to a modification of the zeolite sample.

The results arose because, in the apparatus used, (1) a temperature difference between the temperature sensor and the solid material was marked when the mass of solid is large, and (2) with large solid masses significant hydrogen concentration gradients could occur within the solid bed, resulting in nonhomogeneous reduction.

These experimental results show that special care must be taken in comparing results when sample sizes are markedly different. For example, Jacobs et al. [31,53] observed two rate processes for the reduction of Cu^{2+} to Cu^+ in Y zeolite by isothermal measurements, but were unable to distinguish two processes by TPR under their conditions (0.8-1g of zeolite, 80 kPa of hydrogen) [49]. However, Gentry et al. [25], using $4\%vH_2$ in N_2 and a sample mass of 200 mg, clearly resolved the two processes.

E. Particle Size Effects

It is one of the strengths of TPR as a "fingerprint" method that in many cases it can differentiate between materials that have been prepared in nearly identical ways.

Some work recently published by Tonge [16] suggest that such differences may be explained in terms of particle size effects. Tonge's analysis was based on a reduction that accords with the contracting sphere model (see Section II.A), i.e., a reduction that obeys the relationship

$$1 - (1 - \alpha)^{1/3} = kt$$

where α is the fraction reduced and t is time.

Considering a contracting cube model, the rate of reaction is expressed as the rate of decrease in volume V of the solid.

$$\frac{dV}{dt} = \frac{d}{dt}(X_o - 2Kt)^3$$

where K is the linear rate of movement of the reaction interface and X_o is the initial side length.

Incorporating the simple Arrhenius equation and the linear heating rate equation [Eqs. (21) and (22)] gives

$$\frac{dV}{dt} = 6x_o{}^2 A \exp[-E/R(T_o + \beta t)] - 24x_o t A^2 \exp[-2E/R$$

$$(T_o + \beta t)] + 24t^2 A^3 \exp[-3E/R(T_o + \beta t)] \tag{41}$$

Tonge used interactive computer programs using Eq. (41) to investigate the effects on $\frac{dV}{dt}$ (proportional to peak height) versus temperature curves of varying X_o and other parameters. The program used a Newton-Rapton iterative method to calculate the time taken for a particle of a given size to disappear so that calculation on that particle could be stopped at that point. The program was adapted to deal with mixtures of different-sized particles. Calculations using monodisperse systems showed that the peak widths and the temperatures of the peak maximums increase by increasing the values of X_o.

Curves generated using two sizes of particles with a size ratio of 1:10 are shown in Fig. 12. It is significant that when the smaller size only contributes 10% to the total amount of the solid (curve C), considerable change to the major peak is apparent.

The simulation was extended by incorporating an approximation of the log-normal distribution usually used to describe particle size distribution in real powder samples. The log-normal distribution is completely defined by the geometric mean size μ_g and the geometric standard deviation σ_g. Simulated TPR profiles of samples having the same geometric mean size but different geometric standard deviations are shown in Fig. 13. As σ_g increases the analysis predicts that the peak maximum moves to a higher temperature.

Tonge attempted to validate his analysis by studying the TPR of the hydrogen reduction of nickel oxide. This reaction is known to be a simple one-step reduction first order in both hydrogen and nickel oxide. It was established by isothermal studies that this reduction obeyed the contracting sphere equations for different samples of nickel oxide prepared by heating samples of the nitrate, oxalate, or formate.

Although all samples prepared give approximate log-normal distributions of particle size, efforts to demonstrate the effects of varying the geometric standard deviation met with partial success (see Fig. 14) only. In many cases experimental TPR peaks were far from smooth.

Convincing evidence was obtained, however, for the dependence of T_m on particle size. A nickel oxide sample produced from the oxalate at 400°C was fractionated by sedimentation in water. The experimental TPR profiles of different fractions are shown in Fig. 15.

1. Mass Transport Limitations

If the kinetic parameters derived from analysis of TPR profiles are to be meaningful, it is necessary to establish that rate processes are not influenced by rates of mass transport in the solid [48,54].

Mass transport processes, both inter- and intraparticle are characterized by low-activation energies and will thus alter the shape of reduction profiles. While total mass transfer control is thus readily distinguished from total kinetic control there is a broad regime where both processes can operate and total rate is governed by equations of the form

$$\frac{1}{r_{total}} = \frac{1}{r_m} + \frac{1}{r_k} \tag{42}$$

where r is the rate of reaction and m and k refer to mass transfer control and kinetic control, respectively.

It is common practice to check experimentally for mass transport limitations by looking for changes in TPR profiles when particle size is varied, although it is necessary to take into account the possibility of changes in TPR profiles due to genuine particle size effects (see above).

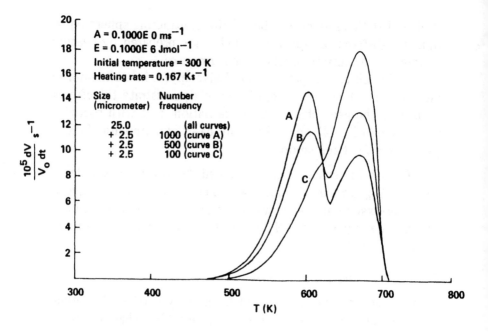

Fig. 12 Simulated TPR profiles of bimodally distributed powders. (From Ref. 16.)

Monti and Baiker [50] established that their TPR results were free from interparticle mass transport limitations by calculating very low values for the concentration gradient between particle surface and the bulk gas phase. They used a correlation by Gunn [55] for interparticle mass transport

$$Sh = \frac{2k_f R_p}{D} = (7 - 10\epsilon + 5\epsilon^2)(1 + 0.7Re_p^{0.2}Sc^{1/3})$$
$$+ (1.33 - 2.4\epsilon + 1.2\epsilon^2)Re_p^{0.7}Sc^{1/3} \qquad (43)$$

where Sh is the Sherwood number, k_f is the fluid-solid mass transfer coefficient, R_p is particle radius, D is the diffusion coefficient, ϵ is the porosity of the sample bed, Re_p is the Reynolds number, and Sc is the Schmidt number.

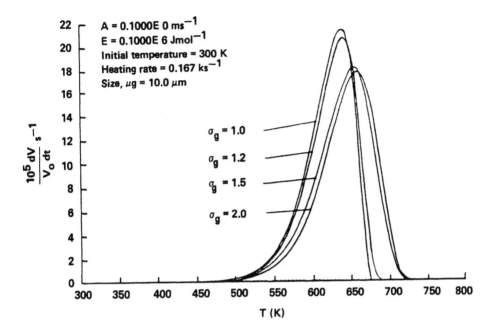

Fig. 13 Effect of varying σ_g on simulated TPR profiles. (From Ref. 16.)

Flows corresponding to Reynolds numbers between 0.06 and 1 were used, and for mass transport by bulk diffusion in a bed of porosity $\epsilon = 0.4$, a limiting value of 3.8 for the Sherwood number was obtained. The concentration gradient between the particle surface and the bulk gas phase was calculated from

$$\left(\frac{dn}{dt}\right)_m = k_f A(c_o - c_s) = \frac{1.9DA}{R_p}(c_o - c_s) \qquad (44)$$

where the outer exchange surface area A was determined by the sample weight and density assuming spherical particles. For reduction of alumina-supported copper oxide results showed values of $(c_o - c_s)/c_o$ were less than 1% (see Table 3).

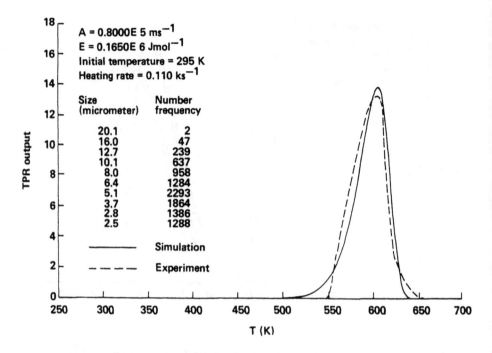

Fig. 14 Comparison of experimental and simulated TPR profiles for a NiO powder. (From Ref. 16.)

For intraparticle diffusion a method presented by Ibok and Ollis [56] for TPD was used by the above authors. The criterion is based on a dimensionless number ϕ defined by

$$\phi = \frac{R_p{}^2}{D_{eff}} \left[\frac{1}{V_c} \left(\frac{dn}{dt} \right)_m \right] \frac{1}{c_s} \tag{45}$$

where V_c is sample volume.

If the values determined for ϕ are less than 0.3 it is concluded that no appreciable diffusion gradients exist within the pellet. For TPR results for the reduction of alumina-supported copper, Monti and Baiker calculated values of ϕ between 0.01 and 0.27 (see Table 3).

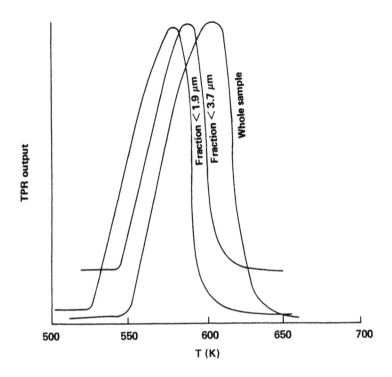

Fig. 15 Effect of particle size on the TPR profile of NiO. (From Ref. 16.)

Heat transfer limitations were considered by Bosch et al. [51] in the TPR of V_2O_5. Dilution of the sample with quartz, however, did not influence the form of the TPR profile.

F. Selection of Values for the Operating Parameters

Certain restrictions on the choice of combinations of operating parameters are self-evident if meaningful TPR profiles are to be obtained. Of particular importance, in the case of hydrogen reduction of an oxide, is that the hydrogen feed rate should be equal to the maximum possible oxygen removal rate; otherwise drastic distortion of the reduction profile will occur. There are

Table 3 Test for Significance of Inter- and Intraparticle
Diffusion Effects in the Reduction of Alumina-Supported
Copper[a]

Experiment	$\overline{R}_p$ (cm)	Re_p (−)	T_M (K)	$\left(\dfrac{dn}{dt}\right)M$ (μmol/s)	ϕ (−)	$\dfrac{c_o - c_s}{c_o} \cdot 10^2$ (−)
31	0.035	0.30	543	1.12	0.27	0.0200
32	0.015	0.10	544	1.17	0.05	0.0037
33	0.0075	0.06	542	1.10	0.01	0.0009

[a]Conditions: $\beta = 0.2$ K s^{-1}; $c_o = 2.46$ μmol cm^{-3}; $V* = 1.25$ cm^3
(NTP) s^{-1}; $S_o = 400$ mg; $K = 127$ s.
Source: Ref. 50.

thus only certain combinations of flow rate, hydrogen concentration, sample mass, and heating rate that are allowable.

Bosch et al. [51] have derived a relation to describe this constraint. Thus,

$$f_{H_2} > r_{red}\, S_{MO_n}$$

where f is the feed rate of hydrogen, r is the specific reduction rate, and S_{MO_n} is the sample weight.

If it is assumed that all the oxygen removed in the particular reduction step is removed at a constant rate during the interval of the peak, $t = \Delta T_{1/2}/\beta$, then

$$S_{MO_n} < f_{H_2}\, \frac{\Delta T_{1/2}}{\beta}\, \frac{M_{MO_n}}{O/M} \tag{46}$$

where

$\Delta T_{1/2}$ is the peak width at half height

O/M is the molar ratio of oxygen removed and metal present

and

MMO_n is the molar weight of MO_n

In order to obtain TPR profiles that can be analyzed using simple kinetic models there are, in fact, two opposing experimental requirements: (1) the hydrogen depletion must be kept as low as possible and (2) enough depletion must occur to ensure reasonable precision of the measurements.

Monti and Baiker [50] propose two criteria to meet these requirements: (1) the amount of hydrogen consumed at the peak maximum should not exceed two-thirds of the hydrogen feed to the reactor and (2) the minimum conversion at the peak maximum should be 10%. Using these criteria for given values of heating rates, Monti and Baiker computed a set of operating variables (sample mass, flow rate, hydrogen concentration, and heating rate) that meet the two criteria.

The molar hydrogen flux leaving the reactor at temperature, T_m, is plotted as a function of the sample mass for a series of heating rates (see Fig. 16). The combination of experimental parameters that satisfy the two criteria are those within the rectangular area shown.

Monti and Baiker have defined a characteristic number, K, as an aid to the selection of appropriate variables

$$K = \frac{S_o}{F[H_2]} \tag{47}$$

They calculated for the hydrogen reduction of nickel oxide (a simple one-step reduction) a minimum value of K of 55 s for a heating rate of $0.1~K~s^{-1}$ and a maximum value of K of 140 s

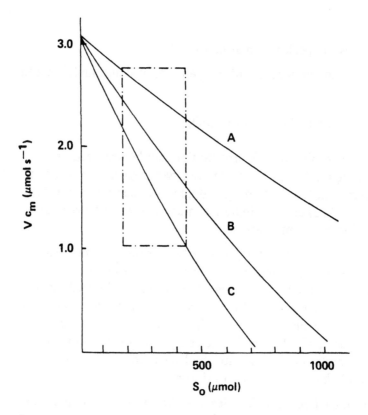

Fig. 16 Minimum molecular hydrogen flow rate at the reactor outlet calculated for different heating rates (A:0.1 K/s; B:0.2 K/s; C:0.3 K/s) and amounts of reducible sample. Conditions: $c_o = 2.46$ μmol/cm^3; F = 1.25 cm^3(NTP)/s. (From Ref. 50.)

for a heating rate of 0.3 K s^{-1}. Thus, for values of K below 55 s the sensitivity becomes too low and at values of K above 140 s the amount of hydrogen consumed becomes too large.

The authors suggest that sets of operating variables can be selected by calculation of the factor K or construction of diagrams shown in Fig. 16.

REFERENCES

1. J. Haber, *J. Less-Common Met.*, *54*: 243 (1977)

2. B. Delmon, *Introduction à la Cinétique Hétérogène*, Edition Technip, Paris (1969)

3. H. W. St. Clair, *Trans. Metall. Soc.*, AIME, *233*: 1145 (1965)

4. B. B. L. Seth and H. U. Ross, *Trans. Metall. Soc.*, AIME, *233*: 180 (1965)

5. R. H. Spitzer, F. S. Manning, and W. O. Philbrook, *Trans. Metall. Soc.*, AIME, *236*: 726 (1966)

6. P. R. Khangaonkar and V. N. Misra, *J. Sci. Ind. Res.*, *35*: 231 (1976)

7. R. Hasegawa, *Trans. Natl. Res. Inst. Metall.*, *20*: 21 (1978)

8. J. Szekely and A. Hastaoglu, *Inst. Min. Metall. Trans.*, *85*: C78 (1976)

9. T. Deb Roy, *Physical Chemistry of Process Metallurgy–Richardson Conference (1973)*, Institute of Mining and Metallurgy, London (1974), p. 85

10. V. C. F. Holm and A. Clark, *J. Catal.*, *11*: 305 (1968)

11. E. J. Nowak and R. M. Koros, *J. Catal.*, *7*: 50 (1967)

12. F. N. Hill and P. W. Selwood, *J. Am. Chem. Soc.*, *71*: 2522 (1949)

13. R. N. Pease and H. S. Taylor, *J. Am. Chem. Soc.*, *43*: 2179 (1921)

14. H. H. Voge and L. T. Atkins, *J. Catal.*, *1*: 171 (1962)

15. J. Bandrowski, C. R. Bickling, K. H. Yang, and O. A. Hougen, *Chem. Eng. Sci.*, *17*: 379 (1962)

16. K. H. Tonge, *Thermochimica Acta*, *74*: 151 (1984)

17. H. Charcosset and B. Delmon, *Ind. Chim. Belg.*, *38*: 481 (1973)

18. A. Roman and B. Delmon, *Acad. Sci., Paris, t273*: 1310 (1971)

19. E. Ruchenstein and B. Pulvermacher, *J. Catal.*, *29*: 224 (1973)

20. R. Kramer and M. Andre, *J. Catal.*, *58*: 287 (1979)

21. J. A. Rabo, R. D. Bezman, and M. L. Poutsma, *Acta Phys. Chem.*, *24*: 39 (1978)

22. Kh. M. Minachev and Ya. I. Isakov, in *Zeolite Chemistry and Catalysis* (J. A. Rabo, Ed.), ACS monograph (1976), p. 171

23. J. B. Uytterhoeven, *Acta. Phys. Chem.*, *24*: 53 (1978)

24. H. Beyer, P. A. Jacobs, and J. B. Uytterhoeven, *J. Chem. Soc., Faraday I*, *72*: 674 (1976)

25. S. J. Gentry, N. W. Hurst, and A. Jones, *J. Chem. Soc., Faraday I*, *75*: 1688 (1979)

26. H. K. Beyer and P. A. Jacobs, *ACS Symp. Ser.*, *40*: 493 (1977)

27. P. A. Jacobs, H. Nijs, and J. Verdonck, *J. Chem. Soc., Faraday I*, *75*: 1196 (1979)

28. M. Briend-Faure, J. Jeanjean, M. Kermarec, and D. Delafosse, *J. Chem. Soc., Faraday I, 74*: 1538 (1978)

29. E. D. Garbowski and J. C. Vedrine, *Chem. Phys. Lett., 48*: 550 (1977)

30. E. D. Garbowski and M. Primet, *Chem. Phys. Lett., 49*: 247 (1977)

31. P. A. Jacobs, M. Tielen, J. P. Linart, J. B. Uytterhoeven, and H. Beyer, *J. Chem. Soc., Paraday I, 72*: 2793 (1976)

32. I. R. Leith and H. F. Leach, *Proc. R. Soc. A., 330*: 247 (1972)

33. C. Kemball, H. F. Leach, and I. R. Leith, *J. Chem. Res., 4*: (1977)

34. J. Texter, D. H. Strome, R. G. Herman, and K. Kleir, *J. Phys. Chem., 81*: 333 (1977)

35. J. M. Zembla, P. Grange, and B. Delmon, *C. R. Acad. Sci., 561*: 279 (1974)

36. B. Delmon, *Proceedings of the International Symposium, 7th (1972) on the Reactivity of Solids*, Chapman and Hall, London (1972)

37. G. E. Batley, A. Ekstrom, and D. A. Johnson, *J. Catal., 34*: 368 (1974)

38. N. I. Il'Chenko, *Russ. Chem. Rev., 41*: 47 (1972)

39. M. Bulens, *Ann. Chim., tI*: 13 (1976)

40. S. J. Gentry, N. W. Hurst, and A. Jones, *J. Chem. Soc., Faraday I, 77*: 603 (1981)

41. H. Charcosset, R. Fretz, A. Soldat, and Y. Trambouze, *J. Catal.*, *22*: 204 (1971)

42. R. C. Mackenzie (Ed.), *Differential Thermal Analysis*, Academic Press, London and New York (1970)

43. J. L. Falconer and J. A. Schwarz, *Catal. Rev.—Sci. and Eng.*, *25* (2): 141 (1983)

44. A. K. Banarjee, S. R. Naidu, N. C. Ganguli, and S. P. Sen, *Technology*, *11*: 249 (1974)

45. A. K. Banerjee, *Technology*, *11*: 162 (1974)

46. N. W. Coats and J. P. Redfern, *Nature*, *201*: 68 (1964)

47. E. D. Rainville, *Special Functions*, Macmillan, New York (1960), p. 44

48. J. M. Thomas and W. J. Thomas, *Introduction to the Principles of Heterogeneous Catalysis*, Academic, London (1967)

49. P. A. Jacobs, J.-P. Linart, H. Nijs, and J. B. Uytterhoeven, *J. Chem. Soc., Faraday I*, *73*: 1745 (1977)

50. D. A. M. Monti and A. Baiker, *J. Catal.*, *83*: 323 (1983)

51. H. Bosch, B. J. Kip, J. C. van Ommen, and P. J. Gellings, *J. Chem. Soc. Faraday I*, *80*: 2479 (1984)

52. R. J. Cvetanovic and Y. Amenomiya, in *Advances in Catalysis* (D. D. Ely, H. Pines, and P. B. Weisz, Eds.), *17*: 103 (1967)

53. R. G. Herman, H. Beyer, P. A. Jacobs, and J. B. Uytterhoeven, *J. Phys. Chem.*, *79*: 2388 (1975)

54. C. N. Sattersfield and T. K. Sherwood, *The Role of Diffusion in Catalysis*, Addison-Wesley, New York (1963)

55. D. J. Gunn, *Int. J. Heat Mass Transfer*, *21*: 467 (1978)

56. E. E. Ibok and D. F. Ollis, *J. Catal.*, *66*: 391 (1980)

3

Experimental Methods and Instrumentation

I. GENERAL PROCEDURE

At the beginning of a TPR experiment, reducing gas is made to flow over a fixed amount of solid at a temperature low enough to prevent reaction. The temperature of the solid is then increased at a linear rate, β K min^{-1}, and the rate of the reaction is monitored. This can be achieved by measuring concentration or pressure changes in the gas phase (reactants or products) or weight changes of the solid.

At the conclusion of the experiment the apparatus must be calibrated in order to obtain quantitative data on the extent of reduction of the various species being studied. This is readily achieved by injecting into the TPR system known amounts of hydrogen, for example, usually at least three different amounts.

The linear calibration curve achieved is then used for the calculation of the extent of reduction of the actual sample under study.

All the TPR experiments in this review have used hydrogen or hydrogen mixtures as the reducing gas. In principle, however, any reducing gas can be used. Furthermore, the technique may be extended to include temperature-programmed reactions in general, e.g., temperature-programmed oxidation [1,2] or temperature-programmed carburization [3], as discussed in more detail in Chapter 5. It is convenient to divide the different experimental methods used in terms of the method of monitoring the reduction.

II. THERMAL CONDUCTIVITY DETECTORS

Because hydrogen has been so commonly used for the reduction, it has been convenient to measure the uptake of hydrogen by the difference in the thermal conductivity of the gas before and after reduction [1,4,5,6,7]. This has been achieved by using low concentrations of hydrogen in nitrogen or argon. Because of its more inert character argon is preferred. Figure 1 shows a typical system [5].

A weighed amount of the sample is placed in the reactor, which is connected to the apparatus. Before the actual TPR measurements the sample may be subjected to a variety of pretreatments using the gas-handling system and a furnace. During TPR the gas stream consists of a low concentration of hydrogen in nitrogen or argon, e.g., 5%v H_2/Ar. This passes through the system at a flow rate of usually 600-1200 ml h^{-1} and an overpressure of 0-0.5 atm, although a wide range of flow rates could be used and in principle the TPR experiment could be conducted under high pressure in a suitably designed system. The reducing gas passes through a deoxygenation catalyst, then a cold trap, through one arm of the thermal conductivity cell, and then through the reactor, which is heated at a linear programmed rate (usually 1-20 K min^{-1}), and thence via a cold

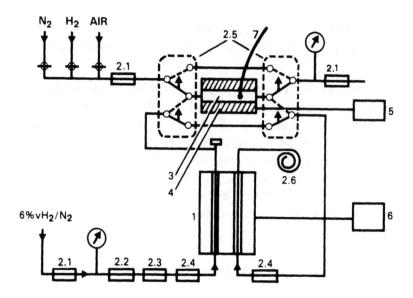

Fig. 1 TPR apparatus using thermal conductivity detection. 1, Thermal conductivity cell. 2.1, Zamba Negretti reduction valve. 2.2, Pt/Al$_2$O$_3$ catalyst. 2.3, Molecular sieves (Linde 5A). 2.4, Dewar (193 K). 2.5, Gas flow switch. 2.6, Brake capillary. 3, Reactor. 4, Oven or furnace. 5, Temperature programmer. 6, Chart recorder. 7, Thermocouple.

trap containing molecular sieves or other adsorbent (to remove reduction products) through the other arm of the thermal conductivity cell where any change in the hydrogen concentration is monitored. The change in hydrogen concentration with time is displayed on the recorder. Since the gas flow is constant, the change in hydrogen concentration is proportional to the rate of reduction. Distinct reduction processes in the sample show up as peaks in the TPR profile.

III. DETECTION BY PRESSURE CHANGES OF THE REACTING GAS

Jacobs et al. [9] have studied the reduction of transition metal ion-exchanged zeolites using 100% hydrogen, and oxidation, using 100% oxygen, in a recirculation reactor schematically represented in Fig. 2.

A microprocessor is used for completely automated TPR, TPO, and TPD experiments. Several temperature profiles can be used with the air-stirred furnace that contains the reactor. The temperature of the catalyst is registered continuously in the computer. At the same time, the pressure in the recirculation reactor is measured with a pressure transducer. Because of pump pulsations this cannot be done continuously, but the interval at which the pressure is registered can be programmed. Before each pressure measurement the circulation pump is stopped for a given amount of time (which also is variable) till the pressure reading is constant.

The total reactor has an internal volume of about 80 ml, and manual calibration with helium is done before each run. The desk computer stores the readings of the reaction temperature and the system pressure. When the run is terminated, the amount of gas uptake (or desorption) is calculated, and, by using the necessary algorithms, the corresponding rates are determined.

Holm and Clark [8] used a similar system to study reduction of supported metal oxides. This work was not strictly TPR since rates were measured at a number of fixed temperatures to obtain reduction profiles. The apparatus consisted of a closed circulating system including a magnetic pump, a furnace containing a "hairpin" reactor for the catalyst, a tube of anhydrous magnesium perchlorate for water removal, and a full-length manometer made of 2-mm i.d. capillary tubing for measuring pressure changes. The circulating system formed a hydrogen reservoir of about 100-ml capacity at atmospheric pressure, and samples of catalysts were taken that required about 20 to 40 ml of hydrogen

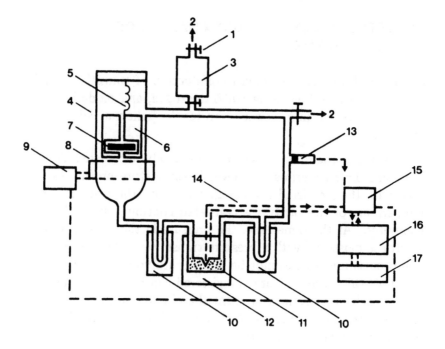

Fig. 2 Recirculation reactor for TPR, TPO, and TPD experiments, 1, Stopcock. 2, To vacuum and gas handling system. 3, Calibrated volume. 4, Circulation pump with variable flow rate (up to 2.5 L min⁻¹). 5, Spring. 6, Piston. 7, Valve. 8, Solenoid. 9, Pump control. 10, Cold trap (198 K). 11, Reactor. 12, Stirred air furnace. 13, Pressure transducer. 14, Thermocouple. 15, Microprocessor. 16, Desk computer. 17, Printerplotter.

for reduction, usually 2 g. A thermocouple placed in the furnace with the catalyst tube was used for temperature measurements. Usually a temperature of 773 K was adequate to obtain complete reduction, but for most coprecipitated catalysts and for impregnated NiO-Al_2O_3 a temperature of 823 K was necessary. Reduction periods ranged from 5 to 24 h. In each case the initial hydrogen pressure was about 1 atm.

IV. DETECTION BY WEIGHT CHANGES OF THE REDUCING SOLID

TPR of supported nickel oxide has been studied by monitoring weight changes during the course of reduction [10,11,12]. Banerjee et al. [10] used a weighed amount of the dry sample, which was contained in a glass bucket hanging from a two-section helical quartz spring enclosed in an all-glass housing. The changes in weight during reduction were observed through a micrometer microscope. The sample was heated by passing a current through a coil wound noninductively over the sample jacket. The rate of temperature rise was maintained at 5 K min^{-1} by a calibrated autotransformer. Electrolytic hydrogen was purified by passing it through an inlet at the top of the bucket. The steam formed during the reaction was carried over by the exit gas through an outlet at the bottom of the sample chamber.

V. OTHER METHODS

We are not aware of examples where TPR has been carried out by monitoring the concentration of the gases produced, or where nonlinear heating schedules have been used, but see the article listed in Ref. 13.

VI. FURNACE DESIGN

TPR may be carried out using any suitable tube furnace or stirred-air oven. Figure 3 shows a tube furnace of low mass

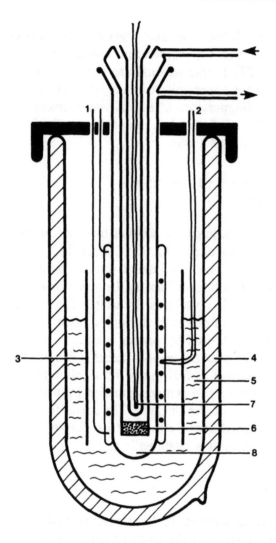

Fig. 3   Low-temperature apparatus for TPR.   1, Insulated ni-chrome windings.   2, Control thermocouple.   3, Nickel foil. 4, Dewar.   5, Coolant.   6, Sinter.   7, Monitoring thermocouple. 8, Reactor.

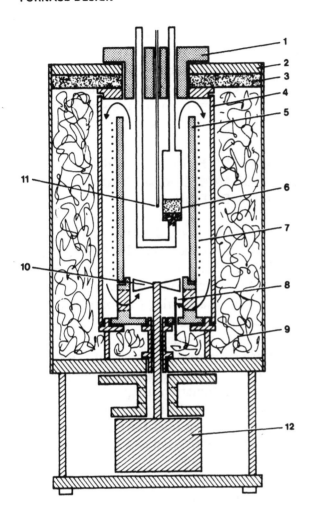

Fig. 4 Stirred air furnace for TPR. 1, Ceramic cover plate.
2, Oven cover. 3, Insulating plate. 4, Stainless steel tube. 5,
Ceramic tube. 6, Reactor. 7, Heating wire. 8, 11, Thermo-
couples. 9, Insulating material. 10, Fan. 12, Motor.

supplied by Stanton Redcroft Ltd. The tube furnace may be immersed in liquid N_2 or surrounded by solid CO_2 to achieve linear temperature programming from subambient temperatures. The stirred-air oven shown in Fig. 4 has an efficient internal air circulation system that prevents overheating of the sample by the heat of reaction.

VII.   AUTOMATED APPARATUS FOR MEASURING TEMPERATURE-PROGRAMMED REDUCTION, DESORPTION, AND OXIDATION

Recently a comprehensive paper by Boer, Boersma, and Wagstaff appeared [13] that described an apparatus in which TPR, TPO, and TPD could be carried out. This was an automated apparatus in which the solid sample under study could be heated or cooled under temperature programming in a flowing gas containing either hydrogen or oxygen as the reactive component using thermal conductivity detection. In this work the design considerations and automation scheme are discussed in depth. Details of the oven and its temperature control, sample holder, gas dosing circuit, assembly of the apparatus, and automation using a Universal Programming Unit are fully described. This is the most comprehensive description to date of a system used for the carrying out of temperature-programmed characterization and is recommended reading for the initiate to the technique.

VIII.   PRACTICAL CONSIDERATIONS

As a penultimate section here it is of value to record some practical considerations when setting out to do TPR experiments. The parameters that have to be optimized are:

Flow rate

Gas ratio

Sample volume/mass

Particle size

Geometry of the bed

Dimensions of the reaction vessel

Heating rate

Signal intensity

A. Flow Rate

If the period between the reduction of the sample and the detection of hydrogen is significantly long, then measured temperatures require correction according to

$$T = T_R - \frac{\beta V}{F}$$

where T_R is the measured temperature, V is the dead volume between reactor and detector, and F is the flow rate. If such corrections are large, the precision of the temperature measurement can be reduced and it is thus preferable that the flow rate be maintained high. The flow rate must also be compatible with the geometry of the reaction vessel such that the gas flow maintains uniform velocity distribution conditions. The flow rate determines the size of the pipework used. Diffusion in the pipes can cause loss of definition if the time gap between reaction and detection is too long (see Chapter 2, Section IV.C.1).

B. Gas Ratio

To avoid distortion of TPR peaks, it is necessary that the hydrogen consumed at maximum reduction be kept to a minimum (see Chapter 2, Section IV.F). The characteristics of the hydrogen/inert gas should be considered so that advantage is taken of the maximum difference in thermal conductivity. Also, the gas ratio should allow use of the linear part of the thermal conductivity plot.

C. Sample Volume/Mass

The volume and mass of the sample are best kept to a minimum as this diminishes problems due to back pressure, temperature differences between the temperature sensor and the sample, and nonuniform gas concentrations within the solid. The smaller amount of gas used is therefore easier to handle, and less water is formed, thus avoiding problems in the water removal system (see also Chapter 2, Section IV.D).

D. Sample Particle Size

Particles that are too small can result in the buildup of back pressure and consequent problems in maintaining constant flow. Particles that are too large result in intraparticle diffusion problems, causing distorted peaks (see Chapter 2, IV.E.1).

E. Geometry of the Reaction Vessel

This determines the sample bed dimensions. Too thin a layer results in an irregular bed. Too deep a bed results in back pressure and flow changes. Probably the best arrangement is to have roughly equal depth and width. A long bed also requires that the temperature plateau of the furnace be long.

F. Heating Rate

There are advantages in having the heating rate as fast as other factors allow. The result is better defined-peaks, less time per run, and less time for changes in the flow rate and base line. However, the temperature programmer must be able to cope to maintain a linear profile and problems can arise, such as diffusion limitations giving distorted peaks if the heating rate is too high. (See also Chapter 2, IV.A.)

G. Signal Intensity

Small peaks need to be adequately defined without undue noise interference. Large peaks are problematical if changes in attenuation are necessary. The signal intensity should be such that unacceptable changes in flow rate show up as a change in the base line if continuous monitoring of the flow rate is not carried out.

IX. CONCLUSIONS

To date, with the possible exception of the work of Boer et al. [13], most experimental apparatus for carrying out TPR has been put together using parts from chromatographic equipment or has been built in the laboratory using a variety of furnaces, and specially constructed ancillary apparatus. So far no equipment manufacturer has produced a black box that will measure TPR. This is perhaps due to the fact that existing devices for measuring, say, BET surface areas such as Perkin Elmer Sorptometers or Thermal Analysis equipment can be fairly easily modified to perform TPR. Nevertheless we foresee in the not-too-distant future the manufacture of a temperature-programmed reaction instrument that will be fully automated and easily changed from reduction mode to, for example, oxidation or sulphidation or desorption.

All experimental apparatus used at present operates at ambient or subambient pressures. There is no reason why high-pressure TPR devices could not be made, and since some TPR results at high pressures will have great technical value, we foresee too that this development will take place soon.

REFERENCES

1. S. D. Robertson, B. D. McNicol, J. H. de Baas, S. C. Kloet, and J. W. Jenkins, *J. Catal.*, *37*: 424 (1975)

2. T. Uda, T. T. Lin, and G. W. Keulks, *J. Catal.*, *62*: 26 (1980)

3. E. E. Unmuth, L. H. Schwartz, and J. B. Butt, *J. Catal.*, *63*: 404 (1980)

4. F. Mahoney, R. Rudham, and J. V. Summers, *J. Chem. Soc., Faraday I*, 75: 314 (1979)

5. B. D. McNicol, in *Proc. Conf. on Characterization of Catalysts* (R. Lambert and J. M. Thomas, Eds.), John Wiley, Chichester (1980)

6. S. J. Gentry, N. W. Hurst, and A. Jones, *J. Chem. Soc., Faraday I*, 75: 1688 (1979)

7. T. Paryjczak, J. Rynkowski, and S. Karski, *J. Chromatogr.*, 188: 254 (1980)

8. V. D. F. Holm and A. Clark, *J. Catal.*, 11: 305 (1968)

9. P. A. Jacobs, M. Tielen, J. P. Linart, J. B. Uytterhoeven, and H. Beyer, *J. Chem. Soc., Faraday I*, 72: 2793 (1976)

10. A. K. Banerjee, S. R. Naidu, N. C. Ganguli, and S. P. Sen, *Technology*, 11: 249 (1974)

11. A. K. Banerjee, *Technology*, 11: 162 (1974)

12. E. E. Unmuth, L. H. Schwartz, and J. B. Butt, *J. Catal.*, 61: 242 (1980)

13. H. Boer, W. J. Boersma, and N. Wagstaff, *Rev. Sci. Instrum.*, 53: 349 (1982)

4

Applications

I. ZEOLITES

Zeolites are crystalline aluminosilicates built up from a framework of silicon and aluminium ions interconnected via oxygen ions. The silicon and aluminium ions are tetrahedrally coordinated to oxygen atoms to produce a three-dimensional aluminosilicate network that results in the formation of channels or cavities in which are located hydrated alkali metal cations balancing the negative charge on each aluminium-oxygen tetrahedron. An example of this is shown in Fig. 1, where the three-dimensional diamond-type structure of the zeolite faujasite is clearly seen.

Because of the molecular dimensions of the cavities and their historic use as separating and drying agents, the term *molecular sieve* has been given to zeolites. The alkali metal ions

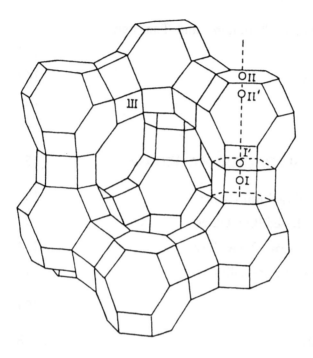

Fig. 1 Faujasite framework and cation siting. Roman numerals indicate cation sites; apices represent Si or Al atoms; oxygen atoms are at the midpoint between apices.

can usually be exchanged for other cations by simply shaking or heating the zeolite with an appropriate aqueous solution of the salt containing the metal to be exchanged on the zeolite. More recently zeolites have found extensive use as catalysts, particularly in petroleum processing. In the case of a catalytic application the metal salt might be a transition metal salt or it might be an ammonium salt. The exchanged zeolite is subjected to drying and calcining to generate a framework containing surface hydroxyl groups that are acidic and active for carbonium ion reactions. The transition metal, which may be used to provide a hydrogenation or other function in the catalyst, is then reduced to the metallic state.

When the zeolite is being used as a separating agent, similar ion exchange and drying procedures are used to modify the size of the entrance ports to the zeolitic cavities.

Clearly, it is important to know the positions of the zeolitic cations in the framework to understand the catalytic and separating functions. It has proved difficult to synthesize zeolites of sufficient single-crystal size for pinpointing cation positions reliably by X-ray single-crystal diffraction. TPR, in conjunction with other techniques, has enabled us both to characterize the state of the cations and, in some cases, to say something about their location in the zeolite, which can be crucial to their catalytic or separating function. Some examples from the recent literature that will illustrate this point are discussed below.

A. Chromium

The TPR profiles of (Cr^{3+}, Na)-X-20 (notation X zeolite 20% Na^+ exchanged with Cr^{3+}) after nitrogen and oxygen pretreatments have been reported by Mahoney et al. [1] (Fig. 2). Following pretreatment in oxygen, the total hydrogen consumed corresponded to 1.1 mol H_2 per mol of exchanged ion. Since 1.5 mol of H_2 per mol of Cr^{3+} would be required for complete reduction to Cr^0, the three peaks observed in the TPR profile may correspond to the reduction steps $Cr^{3+} \rightarrow Cr^{2+}$, $Cr^{2+} \rightarrow Cr^+$, and $Cr^+ \rightarrow Cr^0$. The zeolite framework would seem to stabilize the Cr^+ oxidation state. Pretreatment in nitrogen causes strong autoreduction of $Cr^{3+} \rightarrow Cr^{2+}$ and $Cr^{2+} \rightarrow Cr^+$ (loss of intensity in the first two processes) but does not reduce the Cr^+ to the metal.

Wichterlova et al. [2] in a study of CrY zeolite showed by TPR that Cr^{3+} was reduced to the Cr^{2+} state in the temperature range 570–870 K. On reoxidation, Cr^{2+} was oxidized to Cr^{5+} and Cr^{6+}, and the net valence change upon subsequent reduction was 4.5. These workers suggested that reduction to a state lower than Cr^{2+} was thermodynamically improbable. They also showed that Cr^{3+} was not autoreduced by vacuum treatment.

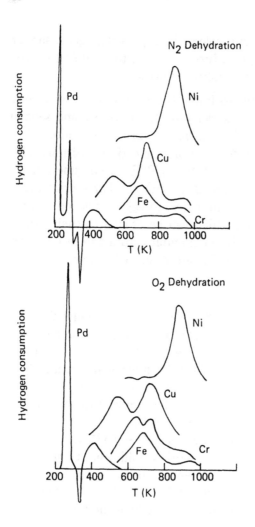

Fig. 2 TPR profiles for (M^{n+}, Na)-X-20 $(M^{n+} = Pd^{2+}, Cu^{2+}, Fe^{2+}, Cr^{3+}, Ni^{2+})$ after dehydration at 723 K in nitrogen or oxygen. (From Ref. 1.)

Thus, in both X and Y zeolites the Cr species is not completely reduced to the metallic state.

B. Manganese

In a study of various cation-exchanged forms of zeolite X [1], no reduction peaks were observed for Mn-containing zeolite.

C. Iron

A dehydrated Fe^{2+} exchanged form of NaX zeolite showed two TPR peaks, one at about 703 K and the other, a very weak peak, at about 923 K (Fig. 2) [1]. However, the hydrogen consumption was much less than that required for even a one-electron reduction. The most likely explanation for this is that a small part of the Fe species is reduced to the zero valent state with the remainder unreduced in inaccessible cation sites.

D. Nickel

The reduction of Ni^{2+} ions within the pores of zeolites has been studied by two groups [1; N. W. Hurst, unpublished results]. For (Ni^{2+}, Na)-X-20 Mahoney et al. [1] observed a single-reduction process for both oxygen and nitrogen pretreated samples, which corresponded to 83 and 91% reduction of $Ni^{2+} \rightarrow Ni^{0}$, respectively (Fig. 2). They found no evidence of a separate $Ni^{2+} \rightarrow Ni^{+}$ reduction step. Hurst [unpublished results] (Fig. 3) found similar results under comparable conditions for (Ni^{2+}, Na)-X-72. Figure 3 also illustrates the influence of sample mass on the TPR profiles of cation-exchanged zeolites. The two reductions observed by Hurst below 1000 K may be attributed to $Ni^{2+} \rightarrow Ni^{+}$ reductions in supercage and sodalite sites. Above 1000 K the reduction continues with the formation of Ni^{0} and possibly the reduction of Ni^{2+} in hexagonal prism sites, although the results of Mahoney and colleagues may suggest that these cations are not reduced. The Ni^{2+} in supercage sites is largely reduced to Ni^{+} by pretreatment in nitrogen.

Jacobs et al. [3] have reported the TPR profile for (Ni^{2+}, Na)-Y-68 for a sample pretreated in vacuum at 773 K for 2 h.

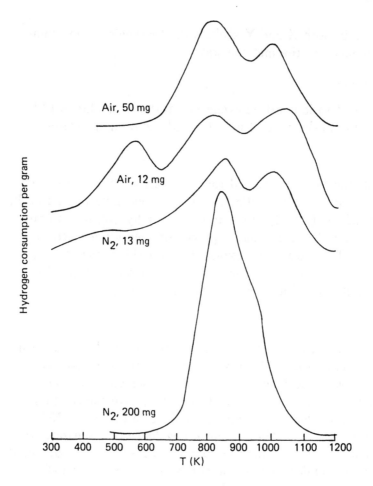

Fig. 3 TPR profiles for different masses of (Ni^{2+}, Na)-X-72 pre-treated at 773 K in nitrogen. (From N. W. Hurst, unpublished results.)

They observed a single peak centered on 773 K. The presence of Ni^0 was demonstrated after the reduction, although the hydrogen consumed only corresponded to about 40% reduction of $Ni^{2+} \rightarrow Ni^0$. They found no evidence of a $Ni^{2+} \rightarrow Ni^+$ reduction process, but the use of low masses and low hydrogen concentrations may distinguish a $Ni^{2+} \rightarrow Ni^+$ process. On reoxidation and subsequent reduction, Jacobs et al. observed the separate low-temperature reduction of $NiO \rightarrow Ni^0$ external to the zeolite framework in the same manner as that observed by Gentry et al. [4] for $CuO \rightarrow Cu^0$ in reoxidized samples of (Cu,Na)-X-50.

E. Copper

The reduction of Cu^{2+} ions within the pores of zeolites has been studied by several groups [1,3,4]. For $(Cu^{2+}$,Na)-X-50 Gentry et al. [4] observed two reduction processes (Fig. 4) which were assigned to $Cu^{2+} \rightarrow Cu^+$ reductions in supercage (550 K) and sodalite cage (650 K) positions. Some Cu^0 was formed during the reduction, but this reduction was not observed as a separate process.

Pretreatment in nitrogen produced much less reduction for both sites than was observed for air-pretreated samples. For $(Cu^{2+}$, Na)-X-20 Mahoney et al. [1] observed two reduction processes assigned to $Cu^{2+} \rightarrow Cu^+$ and $Cu^+ \rightarrow Cu^0$ reductions (Fig. 2). They found no evidence of two processes in the $Cu^{2+} \rightarrow Cu^+$ stage observed by Gentry et al., nor of the fast and slow reduction processes observed for $Cu^{2+} \rightarrow Cu^+$ in $(Cu^{2+}$, Na)-Y-68 [3]. Nitrogen pretreatment produced less reduction in the $Cu^{2+} \rightarrow Cu^+$ stage, indicative of autoreduction of Cu^{2+} ions.

To rationalize these data requires consideration of the conditions of the experiments. Gentry et al. [4] have shown that increasing the mass of $(Cu^{2+}$, Na)-X-50 to 400 mg results in complete loss of resolution of the two $Cu^{2+} \rightarrow Cu^+$ reduction processes in the TPR profile. The use of a 1.0-g sample by Mahoney et al. [1] resulted in the observation of a single $Cu^{2+} \rightarrow Cu^+$ reduction process.

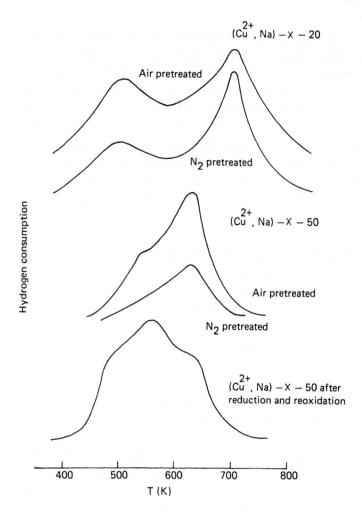

Fig. 4 TPR profiles for (Cu^{2+}, Na)-X. (From Ref. 4.)

The observation by Mahoney et al. of a $Cu^+ \rightarrow Cu^o$ step not observed by Gentry et al. may be explained by the higher hydrogen concentration, lower heating rate, and finally the extended temperature range of the experiment.

For (Cu^{2+}, Na)-Y-68 Jacobs et al. observed a TPR profile with two processes assigned to $Cu^{2+} \rightarrow Cu^+$ and $Cu^+ \rightarrow Cu^o$ reductions [3]. The conditions used by Jacobs et al. were similar to those used by Mahoney et al., i.e., 1.0-g sample and a heating rate of 5 K min^{-1}, but the experiments were carried out in a recirculating system with a starting pressure of 80 kN m^{-2} of pure hydrogen. In contrast to this, Gentry et al. [4], as mentioned above, using the mild conditions described for (Cu^{2+}, Na)-X-50, observed two peaks assigned to $Cu^{2+} \rightarrow Cu^+$ reductions in sodalite and supercage positions (Fig. 4). Using isothermal methods, Jacobs et al. had previously been able to distinguish two processes for the $Cu^{2+} \rightarrow Cu^+$ stage of the reduction and calculated activation energies for these processes as 111 and 70 kJ mol^{-1}, respectively [5]. Gentry et al. obtained values of 84 and 64 kJ mol^{-1}, respectively, for these processes using TPR. Pretreatment in nitrogen resulted in less reduction during the low-temperature process observed by Gentry, indicative of autoreduction of cations occupying supercage positions.

Reoxidation has been studied by Gentry et al. and also by Jacobs et al. Gentry et al. found that for (Cu^{2+}, Na)-Y-68, the Cu^+ ions formed by gentle reduction were completely reoxidized to Cu^{2+} ions occupying their original zeolite lattice positions. However, for (Cu^{2+}, Na)-X-50 at least some of the Cu^o external to the zeolite was reoxidized to CuO external to the lattice which subsequently reduced by a separate process (Fig. 4). For (Cu^{2+}, Na)-Y-68, reduced to Cu^o by more harsh reduction, Jacobs et al. observed in the temperature-programmed oxidation profile two peaks that were assigned to the reoxidations of Cu^o in the zeolite to Cu^{2+} ions in zeolite positions and of Cu^o external to the zeolite framework to CuO. Subsequent TPR failed to show the separate reduction of the CuO although the TPR profile was broader than for a non-reoxidized sample. The failure to observe this process, observed by Gentry et al. for reoxidized (Cu^{2+}, Na)-X-50, may be due to the mass of zeolite used in the work.

F. Ruthenium

The TPR profile for X zeolite about 46% exchanged using $[(NH_3)_6Ru] Cl_3$ has been reported [6] and reveals the existence of three maxima in the rate of hydrogen uptake. The average valency of Ru after each maximum may be represented by

$$Ru^{3+} \longrightarrow Ru^{2+} \longrightarrow Ru^+ \longrightarrow Ru^o$$

$$423\ K \qquad 493\ K \qquad 563\ K$$

It was not possible under the conditions used to distinguish between ruthenium ions in different zeolite locations. A sample of (Ru^{3+}, Na)-Y-40 degassed and reduced at 773 K was completely reoxidized at room temperature to RuO_2. A subsequent TPR experiment caused complete reduction to Ru^o crystallites [7].

G. Palladium

The reducibility of palladium has been studied in zeolites by Mahoney et al. [1]. X zeolites in which some of the Na^+ ions were exchanged for Pd^{2+} ions were dehydrated in oxygen or nitrogen at 723 K prior to TPR. After dehydration in nitrogen, two reduction peaks were observed in the TPR profile at 210 and 240 K followed by two hydrogen desorption peaks at 320 and 340 K (Fig. 2). A further reduction peak somewhat broader than the low-temperature peaks occurs at 410 K.

The TPR profile obtained when the zeolite was dehydrated in oxygen rather than nitrogen was less complicated, showing a single hydrogen uptake around 270 K and desorption around 340 K, followed by a broader hydrogen uptake around 410 K.

TPR profiles obtained by Hurst [unpublished results] using Y zeolite that had been ion exchanged using $[(NH_3)_4Pd]Cl_2$ at

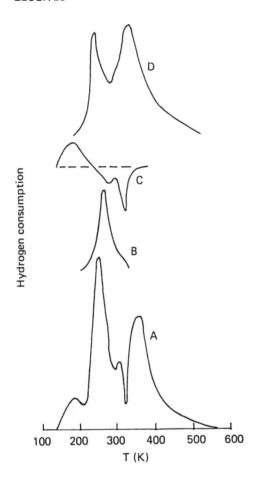

Fig. 5 TPR and TPD profiles of Pd^{2+} exchanged Y zeolite.
(A) TPR profile. (B) Hydrogen absorption curve. (C) TPD
profile. (D) Resultant TPR profile. (From N. W. Hurst, unpublished results.)

pH = 6.0 are in agreement with these results. The TPR profile
obtained (β =15 K/min^{-1}, 10% H$_2$/N$_2$, 20 ml/min^{-1}), using a 50-mg
sample following pretreatment in air to 775 K, is shown in Fig.
5, curve A. After the TPR experiment the sample was cooled
in flowing 10% H$_2$/N$_2$ (curve B). A subsequent TPD experiment
using 10% H$_2$/N$_2$ produced curve C. Curve D is the resultant
profile produced by the subtraction of curves B and C from
curve A. Curve B represents the hydrogen adsorption component
of curve A, while curve C represents the desorption component
(the apparent hydrogen uptake around 150 K is nitrogen desorption from the zeolite). Thus, curve D represents the TPR profile
for the reduction of Pd^{2+} ions within the pores of Y zeolite with
adsorption/desorption effects removed. It may be interpreted as
Pd^{2+} → Pd^o reductions of Pd^{2+} ions in two different zeolite locations (supercage and sodalite cage). The alternative of a two-
stage reduction Pd^{2+} → Pd$^+$ and Pd$^+$ → Pd^o is not acceptable because Pd^o is present during the first reduction/adsorption process.

II. BULK MONOMETALLIC MATERIALS

Most work on the characterization of materials using TPR has,
as mentioned earlier, been carried out in relation to catalysts,
and understandably this work has been done predominantly on
supported metal catalysts in the oxidic form, which are the
"real" catalyst systems. However, a significant and increasing
amount of work has been done on bulk solids which has provided valuable information on the ease of reduction, valence
state of metals, metal-metal interactions in bulk mixtures, and
influence of one component in a mixture on the reducibility of
other components. Our approach in reviewing this area is to
deal with each specific material in separate sections according to
the particular metal being studied and then to look at the influence of dopants in a separate section.

A. Iron

Little work has been done on the reducibility of bulk iron oxides by TPR. In a study of reforming catalysts McNicol [unpublished results] examined the reducibility of Fe_2O_3 physically mixed with both pure γ-Al_2O_3 and with a reforming catalyst of composition 0.375%w Pt on γ-Al_2O_3. In both cases it was not expected that the physical mixing produced any chemical interaction between the Fe_2O_3 and the Al_2O_3. The results of this investigation are illustrated in Fig. 6. In the case of Fe_2O_3 mixed with pure γ-Al_2O_3 the Fe_2O_3 was seen to reduce in a series of peaks above 600 K but with a total hydrogen consumption corresponding to about half of that required for the complete reduction of Fe_2O_3 to the metallic state. For the Fe_2O_3 mixed with the Pt/Al_2O_3 the consumption of hydrogen was also about half of that required for reduction of Fe_2O_3 to the metal. However, the reduction that did occur took place at lower temperatures than in the Fe_2O_3/Al_2O_3 system. It was proposed that the Pt metal species acted as an agent to dissociate the hydrogen and allowed the hydrogen atoms to easily reach the Fe_2O_3 phase, thus facilitating the reducibility of that phase.

Brown et al. [8] studied the reducibility of bulk iron oxide materials during a study of the reducibility of ammonia synthesis catalysts. Their study showed that the reducibility of Fe_2O_3 as measured by TPR differed markedly from that of Fe_3O_4. The first step in the reduction of Fe_2O_3 is the reduction of Fe_2O_3 to Fe_3O_4. The TPR profiles confirmed this with the low-temperature peak in the profile of Fe_2O_3 corresponding to this reaction.

B. Nickel

TPR of bulk NiO has been studied by Robertson et al. [9]. They report a single reduction process $Ni^{2+} \rightarrow Ni^{o}$ centered on 600 K (6% H_2/N_2, β = 4.5 K min^{-1}, flow rate 10 ml min^{-1}). Brown et al. [8] similarly found one peak for reduction of bulk

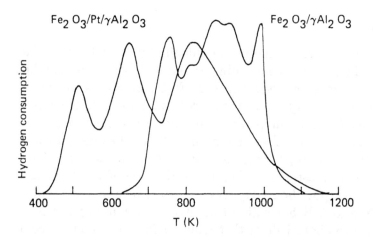

Fig. 6 TPR profiles of Fe_2O_3 physically mixed with γ-Al_2O_3, and Fe_2O_3 physically mixed with (0.375%w Pt, 0.9%w Cl) on γ-Al_2O_3. (From B. D. McNicol, unpublished results.)

NiO at 691 K, some 90 K higher than in the work of Robertson et al. The hydrogen consumption corresponded to reduction of $Ni^{2+} \rightarrow Ni^0$. The discrepancy in reduction temperatures is in part accounted for by the higher heating rate (10 K min^{-1}) and the lower hydrogen concentration (5%v) used by Brown et al. However, details of how the NiO was prepared were not given by Brown et al.

Monti and Baker [10] studied bulk NiO reducibility by TPR and found that it reduced in a temperature region similar to that found by Robertson et al. [9]. They calculated activation energies using the methods outlined in Chapter 2. Their results were in excellent agreement with results reported using conventional methods.

TPR has also been used to study natural garnierite, a nickel-containing mineral [11,12]. The mineral consists of a mixture of a talclike phase and a fiberlike phase. The major amount of nickel is associated with the talclike phase. The TPR profile shows three maxima in the rate of hydrogen uptake, corresponding to three distinct nickel species. The first low-temperature maximum is associated with the talclike phase and corresponds to the reduction of a Ni-oxide-hydroxide phase physically adsorbed on the talclike particles. The broad intermediate maximum of weak intensity is found in a temperature region slightly higher than expected for the reduction of bulk nickel oxide (as determined by reoxidation and a subsequent TPR) and may be due to the reduction of minor quantities of Ni^{2+} present in the fiber phase. The third maximum is observed only when the mineral starts to decompose and may be assigned to the reduction of lattice-substituted Ni^{2+} ions, which are liberated only when the crystallinity of the material declines. This kind of Ni^{2+} is predominantly present in the talclike material.

Similar studies using TPR have been made [13] on Ni hydroxymontmorillonite, and in this case also three reduction peaks were observed. The intensity of each peak was dependent on the nickel content. The high-temperature peak arose from the incorporation and subsequent reduction of Ni ions in the silicate framework. The intensity of this band could be increased by preheating. Work has also been done on Ni-containing serpentine and talc and it was shown how TPR could be used as a fingerprint for such Ni-containing minerals and may help to identify components in a mixture and give a semiquantitative estimation of their concentration.

C. Copper

The TPR of unsupported CuO shows a single reduction peak and has been studied in some detail by Gentry et al. [14]. The effects of heating rate, mass, and hydrogen concentration were investigated for CuO prepared by nitrate calcination and for a sample of Johnson Matthey (JM) CuO. Equation (34) of Chapter 2 was used to determine an overall activation energy for the reduction process. These results are shown in Fig. 7. Although there is some scatter, a single straight line is seen to fit these data, which include the effect of variation of the hydrogen

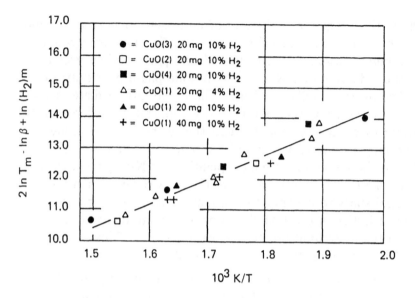

Fig. 7 Activation energy plot for copper (II) oxides. CuO(1) = Johnson Matthey Spec. Pure. CuO(2) = calcined Analar $Cu(NO_3)_2$ $3H_2O$. CuO(3) = calcined $Cu(NO_3)_2$. $3H_2O$ from CuO(1). CuO(4) = calcined $Cu(NO_3)_2$ $3H_2O$ recrystallized. (From Ref. 14.)

concentration in agreement with Schoepp and Hajal [15], who concluded that the reaction is first order with respect to hydrogen concentration. The slope of this line yields an activation energy of $E = 67 \pm 10$ kJ mol^{-1} in good agreement with isothermal results. Robertson et al. [9] also reported the TPR profile for a sample of CuO obtained by nitrate calcination. They observed a single peak centered on 640 K ($\beta = 6.5$ K min^{-1}, 6%v H_2/N_2, flow rate 10 ml min^{-1}). Gentry et al. obtained a value of 553 K for (JM) CuO at the same heating rate but using 10%v H_2/N_2 and a flow rate of 20 ml min^{-1}. In this case the higher flow rate and higher hydrogen concentration may be expected to have lowered T_m by about 70 K [14], so that an adjusted value of T_m might be 623 K. A value of 611 K was also obtained by Gentry for a calcined nitrate sample under conditions similar to those used by Robertson et al. This is a typical spread of values, 611–640 K, for different samples using different apparatuses that should give nominally identical results. However, much greater variation will be observed if the effects of hydrogen concentration, flow rate, and heating rate are not considered.

D. Ruthenium

Ruthenium is catalytically important for a number of reactions, in particular as Ru metal in ammonia synthesis and Fischer-Tropsch hydrocarbon synthesis, and as RuO_2 in the electrocatalysis of the evolution of chlorine. It is also the most effective prometer of platinum for methanol electro-oxidation [16].

There has been some controversy in the literature regarding the reducibility of Ru oxides by electrochemical means and some work has been carried out on the TPR characteristics of Ru-containing catalysts. McNicol and Short [17] examined the reducibility of bulk unsupported Adams RuO_2 catalyst and found that it reduced in a single peak at 443 K. This compares with temperatures of 298 and 348 K for Adams PtO_2, a related electrocatalyst (Fig. 8). The stoichiometry of reduction was that of a four-electron process as expected. The difficulty of reduction of RuO_2 compared with that of PtO_2 is consistent with electrochemical studies of the reduction of these two oxides. Similarly Bossi et al. [18] found that bulk RuO_2 reduced in a peak at 443 K under similar TPR conditions to those used by McNicol and Short [17]. Likewise, the stoichiometry of the reduction was a four-electron process.

Salts of ruthenium such as $RuCl_3$ and $RuNO(NO_3)_x$ are of interest since these materials are generally used in the preparation of ruthenium-containing catalysts. $RuNO(NO_3)_x$ can exist in different forms, with x ranging from 2 to 5 depending on the acidity of the solution from which the salt is prepared. Consequently, TPR profiles of $RuNO(NO_3)_x$ produced from 3 and 12 mol dm^{-3} HNO_3 are significantly different [19]. The hydrogen consumptions are of an order consistent with a reduction reaction

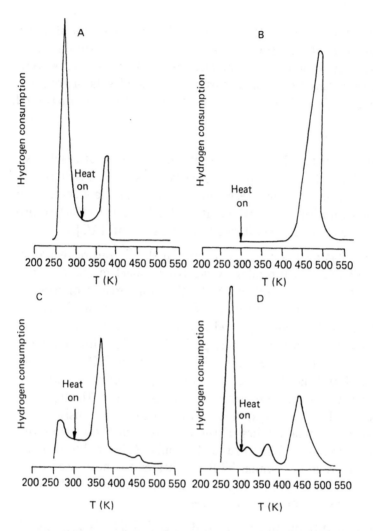

Fig. 8 TPR profiles for platinum and ruthenium Adams oxides. (A) Platinum. (B) Ruthenium. (C) Platinum/ruthenium Adams oxide alloy catalyst. (D) Physical mixture of platinum and ruthenium Adams oxide. (From Ref. 17.)

involving the reduction of the NO and NO_3 groups in the molecule:

$$RuNO(NO_3)_x + \frac{(9x + 5)}{2} H_2 \rightarrow Ru + (x + 1)NH_3$$

$$+ (3x + 1)H_2O$$

E. Tin

Tin is of some interest in oxide form as both catalyst and catalyst support in CO oxidation [20] and as a component in reforming catalysts for increasing the octane rating of petrol [21]. It has recently been shown to act as a promoter of platinum in the electrocatalytic oxidation of methanol [22,23]. The TPR of tin oxides has been studied by Hughes and McNicol [24] in an evaluation of semiconducting tin oxide as a support for platinum in the electro-oxidation of methanol. They found that Analar grade SnO_2 (BDH) is reduced predominantly in a single peak at about 1003 K (β = 6 K min^{-1}, 5%v H_2/N_2, 20 ml min^{-1}) with a hydrogen consumption equivalent to a four-electron reduction. There is also a small but measurable hydrogen consumption between 373 and 773 K. The monoxide (SnO) reduced in a single peak around 1073 K with a hydrogen consumption equivalent to a two-electron process.

TPR profiles obtained by Hurst [unpublished results] using SnO_2 (Hopkins and Williams) are in support of this work. Using 10%v H_2/N_2, 20 ml min^{-1}, β = 15 K min^{-1} produced a TPR profile with a maximum around 1025 K and a small low-temperature uptake of hydrogen around 600 K.

F. Rhenium

Rhenium is catalytically important in hydrogenation/dehydrogenation and cracking reactions and as a second component to platinum in catalytic reforming catalysts. It also acts as a promoter of platinum in methanol electro-oxidation.

Yao and Shelef [25] studied the reducibility of bulk ReO_3 and found that it reduced between 623 and 648 K and consumed an amount of hydrogen equivalent to a six-electron reduction.

G. Vanadium

Catalysts based on vanadium are used for a number of industrial oxidation processes. Recently Bosch et al. [26], Roozeboom et al. [27], and G. C. Bond [unpublished results] have studied the reducibility of vanadium oxides using TPR. Bosch et al. showed that the TPR profile of V_2O_5 under certain conditions consisted of a number of discrete peaks in the temperature range 900–1100 K (Fig. 9). They interpret the results as a stepwise reduction of V_2O_5 to V_2O_3 through intermediates that include V_6O_{13} and VO_2. They also found a high-activation energy for the reduction amounting to about 200 kJ mol^{-1} indicating that solid-state diffusion influences the reduction process for V_2O_5. Bond produced similar TPR results. Conversely, Roozeboom et al. [27] observed the reduction of V_2O_5 to occur in a single peak at 800 K, considerably lower in temperature than found by Bosch et al. [26]. The materials used by both groups were commercial oxide V_2O_5 so the discrepancy can be attributed only to the differing experimental conditions. Roozeboom et al. used the same heating rate as Bosch et al. but used a hydrogen concentration of 66%v, which is very high. Unfortunately, the hydrogen concentration is not given in the work of Bosch et al. but is presumably much lower. If it is indeed lower, then the shift in peak to higher temperatures in Bosch and colleagues' work is in the right direction. However, the resolution into a stepwise reduction found by Bosch et al. yet not observed by Roozeboom et al. is worrying. Given the excellent agreement of TPR studies on other solids among different laboratories where conditions are similar, our conclusion must be that the difference found in the above works must lie with the experimental conditions, although this cannot be unequivocally proved from the information given in the publications. Also, very recently, Monti et al. [28] have shown that the starting temperature

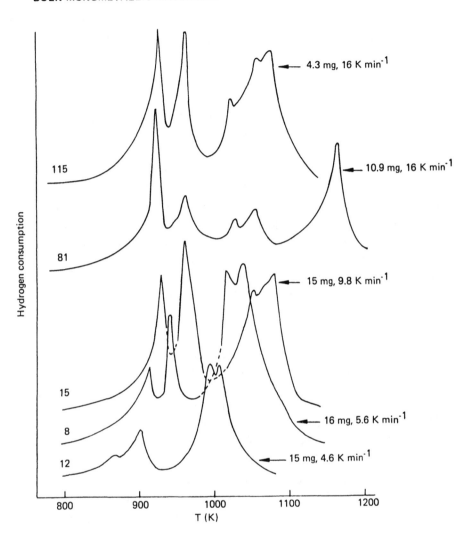

Fig. 9 TPR profiles of V_2O_5—influence of heating rate and sample mass. (From Ref. 26.)

of reduction of V_2O_5 can be reduced by up to 150 K by the addition of 40% K_2SO_4.

H. Cobalt

Cobalt is of interest catalytically as a component in hydrodesulfurization (h.d.s.) catalysts. The reducibility of bulk Co_3O_4 has been studied using TPR [8,29,30,31]. Paryjczak et al. [30] found that the material reduced in a two-stage process with TPR maxima at 593 and 663 K, respectively. This was interpreted as arising from $Co^{3+} \rightarrow Co^{2+}$ in the first peak and $Co^{2+} \rightarrow Co^{\circ}$ in the second peak. Lycourghiotis et al. [29] found that Co_3O_4 reduced at slightly different temperatures. However, both sets of results are almost identical if correction is made for the different conditions used by Lycourghiotis et al., i.e., higher flow rate of 36 ml min^{-1} as opposed to 10 ml min^{-1} and lower hydrogen content 5% v H_2 in Ar as opposed to 8% v H_2 in Ar used by Paryjczak et al.

Bulk Co_3O_4 and CoO were studied by Brown et al. [8] using TPR. The reduction profiles of both oxides were similar showing peaks at around 673 K. The Co_3O_4 sample showed an indication of two peaks at 623 and 723 K possibly corresponding to the reducibilities of the two states of Co in Co_3O_4. The TPR conditions and method of preparation of material were similar to those used by Paryjczak et al. [30].

Recent work by Van't Blik [31] also gave two peaks for reduction of Co_3O_4 with the same interpretation as above. They also studied CoO and found a reduction profile similar to that of Brown et al. [8].

I. Platinum

Platinum is one of the most important metals commercially because of its wide application as a catalytic component for a host of chemical processes. Characterization of the state of platinum in various catalysts has been carried out using photoelectron spectroscopy, X-ray powder diffraction, transmission electron microscopy, chemisorption studies, etc., in order to identify the state of the active platinum component. For many catalysts a plethora of different platinum compounds can be used as a source of platinum. TPR has therefore been carried out [McNicol, unpublished results] on a wide range of platinum compounds of importance in catalyst preparation, e.g., $PtCl_4$, H_2PtCl_6, $Pt(NH_3)_2$ $(NO_2)_2$, $Pt(NH_3)_4(OH_2)$, $H_2Pt(OH)_6$. The profile obtained for samples dried at 393 K for 2 h are shown in Fig. 10. Reduction of $PtCl_4$, $H_2[PtCl_6]$, and $H_2[Pt(OH)_6]$ was by a four-electron process as expected, while reduction of $[Pt(NH_3)_4](OH)_2$ and $[Pt(NH_3)_4]Cl_2$ involved a two-electron process. It is interesting to note that the Cl-containing salts are significantly more difficult to reduce than the OH-containing salts. Also, the TPR profile of $[Pt(NH_3)_4](OH)_2$ displayed four peaks. The origin of this structure is not clear, although the reduction temperature of one of the peaks corresponds to that for $[Pt(NH_3)_4]Cl_2$. $Pt(NH_3)_2(NO_2)_2$ reduced in a single peak at 483 K with a hydrogen consumption that corresponded to the reaction

$$Pt(NH_3)_2(NO_2)_2 + 7H_2 \rightarrow Pt + 4NH_3 + 4H_2O$$

that is, the NO_2 groups in the molecule are reduced. Similar high uptakes of hydrogen have been observed previously in the reduction of other metal nitrates [9].

The TPR of unsupported Adams platinum (IV) oxide has also been studied [17]. The results showed that a two-stage reduction took place—the TPR profile displayed peaks at 298 and 348 K, the former being about twice as intense as the latter (Fig. 8). The hydrogen consumption was equivalent to that of a four-electron reduction. The double peaks may arise from a two-step reduction of Pt^{4+} to Pt^{2+} and then to Pt^o. Yao et al. [32] using TPR found that PtO_2 reduced in a similar temperature range as that found by McNicol and Short [17].

In a study of PtO_2 and RuO_2 catalysts reportedly active for the production of oxygen from water it was discovered

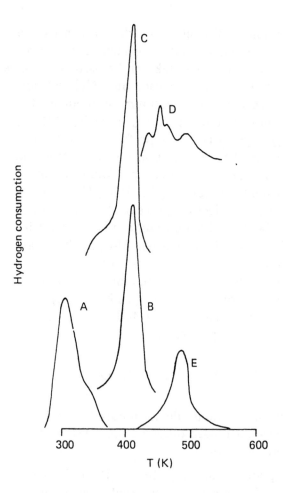

Fig. 10 TPR profiles of some platinum salts used in catalyst preparation. (A) $H_2[Pt(OH)_6]$. (B) H_2PtCl_6. (C) $PtCl_4$. (D) $Pt(NH_3)_4(OH)_2$. (E) $Pt(NH_3)_4Cl_2$. (From B. D. McNicol, unpublished results.)

[P. A. Attwood, G. Brunton, and B. D. McNicol, unpublished results] that the reduced materials that were supposed to be inactive for the reaction showed intrinsic catalytic activity substantially greater than the oxides. TPR studies of the "reduced" catalysts showed reduction peaks for both Pt and Ru at temperatures substantially below that found for the bulk oxides. For "reduced" PtO_2, for instance, a reduction peak was found at 201 K (Fig. 11). From surface area measurements of the "reduced" catalysts it was calculated that the stoichiometries of reduction for the materials corresponded to $PtO_{0.8}$ and RuO_{3-4}. Thus, the reduced catalysts interact with oxygen in the air during transference from reduction furnace to catalytic reactor to produce a surface oxide that is in fact more active than the original bulk oxide catalyst.

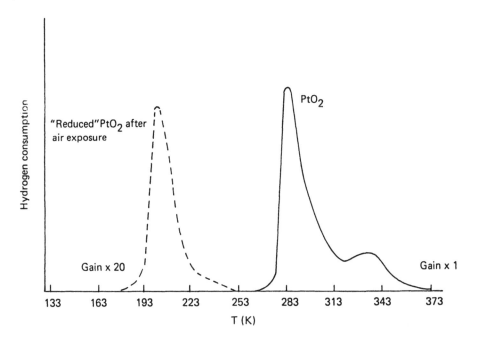

Fig. 11 TPR profiles for platinum Adams (PtO_2) and reduced PtO_2. (From P. A. Attwood, G. Brunton and B. D. McNicol, unpublished results.)

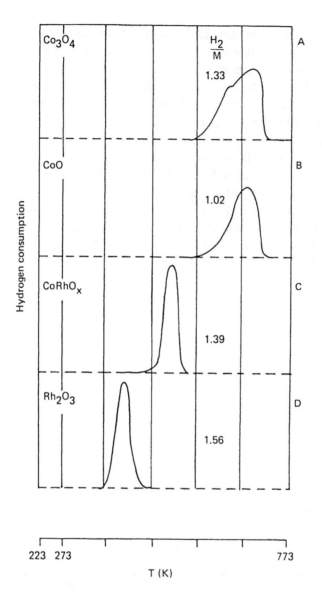

Fig. 12 TPR profiles for bulk oxides. (A) Co_3O_4. (B) CoO. (C) CoRhOx. (D) Rh_2O_3. (From Ref. 31.)

J. Rhodium

Bulk Rh_2O_3 (Made by reduction of $RhCl_3$ at 1073 K followed by oxidation at 1073 K for 72 h) reduced in a single peak at 408 K (Fig. 12) with a hydrogen consumption equivalent to that for complete reduction of Rh^{3+} to the zero valent state [33].

III. BULK BIMETALLIC MATERIALS

Doped metals or bimetallic/multimetallic materials are of increasing importance due to the promoting actions or modifications a second or third component can have on the catalytic activity of a specific metal or on other metallic properties. For example, the formation of alloys between metals can have a marked effect on chemisorption and catalytic properties of metals. It is therefore vital to characterize the state of metallic components in such doped materials to determine if possible the role of the second component in, for example, modifying the role of the primary component.

In this section we will address bulk doped materials by discussing separately each system. A more detailed account is given of copper oxide doped with transition metals since a considerable effort has gone into the characterization of that system using TPR.

A. Platinum/Ruthenium

In the study of methanol electro-oxidation catalysts, unsupported, finely dispersed Pt/Ru catalysts have been shown to be highly active [17]. The activity is thought to arise from an alloy formed between the two metals. The TPR of unsupported Adams-type PtO_2 and RuO_2 has already been discussed (Sections II.D and II.I). The platinum species reduced in two peaks at 298 and 348 K, while the ruthenium species reduced in a single peak at 443 K. A physical mixture of the two oxides showed a TPR profile resembling the sum of the profiles of the two individual metal oxides, implying little or no interaction between the metallic species. As a consequence, the catalytic activity is equivalent to that of the platinum content only. An Adams Pt/Ru oxide catalyst of similar composition to the physical mixture but prepared by the Adams technique as a mixed oxide showed reducibility characteristics intermediate between those of the two bulk oxides, that is, an intense reduction peak at 348 K with weak peaks at 298 and 443 K (Fig. 8). The bulk of the ruthenium was now reducing in the same region as the platinum, and a greater fraction of the platinum was reducing at the higher temperature. This was interpreted as indicating alloy formation between platinum and ruthenium and indeed this was confirmed by X-ray powder diffraction and the electrochemical surface technique of cyclic voltammetry. Furthermore, the catalytic activity for methanol electro-oxidation was about 50 times that of the physical mixture catalyst.

B. Nickel/Cobalt

In a mixed cobalt oxide/nickel oxide catalyst of composition in reduced phase 67% w Co 33% Ni, TPR studies [8] showed a hydrogen uptake corresponding to the cobalt species being present to the extent 64% Co_3O_4. Thus, almost all of the cobalt present exists as Co_3O_4. A mixed oxide of composition 33% w Co 67% w Ni showed a TPR profile corresponding to 40% Co_3O_4. In both cases the amount of hydrogen consumed was greater than that predicted for a physical mixture of the two metal oxides. The results suggest that the presence of nickel helped to stabilize Co_3O_4.

C. Iron/Cobalt

The addition of 5% cobalt to iron was shown by Brown et al. [8] to cause the decrease of the maximum temperature for reduction of Fe_2O_3 from 625 to 603 K and 1009 to 915 K, respectively, indicating that the presence of cobalt facilitates the reduction of iron. This trend in the TPR of the samples was followed as further amounts of Co were added. This behavior was related to the behavior of Fe-Co ammonia synthesis catalysts.

D. Iron/Nickel

TPR of the above system [8] showed that the presence of nickel facilitated the reduction of iron oxides. However, there was evidence from TPR that a spinel phase was formed that did not reduce below 1173 K. Samples containing 30-60%w Fe were almost completely reduced below 873 K, whereas the pure iron oxide was not reduced until temperatures greater than 1173 K were reached. As the iron content of the mixed oxides dropped below 30%w the TPR profiles resembled those of pure NiO.

E. Cobalt/Rhodium

Solid solutions of atomic-ratio Co/Rh = 1 were made by oxidation of evaporated mixed salts of Co and Rh at 1073 K for 48 h [33]. The TPR profile of this material is shown in Fig. 12. The material $CoRhO_x$ reduced in a single peak in the temperature range 453-523 K. The hydrogen consumption was equivalent to that expected for a mixed oxide of composition $CoRh_2O_4$ and Co_3O_4. XRD showed the mixed oxide to consist of these components. The fact that the TPR is a single peak at a temperature lower than that for cobalt oxide indicates that the phase-separated Co_3O_4 is in intimate contact with $CoRh_2O_4$ and that the latter acts as a catalyst for the reduction of the cobalt oxide.

F. Palladium/Vanadium

In a study of the oxidation-reduction properties of palladium/vanadium oxide catalysts using TPR and temperature-programmed oxidation, Wu has shown [34] that a small amount of Pd leads to a significant decrease in the reduction temperature of V_2O_5, and the reduction temperature decreased with increasing Pd content. However the reoxidation temperature did not vary. When the PdO content was close to 0.05%w the reduction and oxidation temperatures are close to each other. This was related to the activity for C_2H_4 oxidation and the regeneration of activity by the oxidation of the reduced catalyst.

G. Copper/Transition Metals

Gentry et al. [14] have made a detailed study of the reducibility of copper oxide, CuO doped separately with 2 mol% of Cr, Mn, Fe, Co, Ni, Ru, Rh, Pd, Ag and Ir, Pt, Au. The notation we shall use here is CuO-2-X where X = Cr, Mn, Fe, etc.

Copper oxide doped with first-row transition metal ions produced TPR profiles that were essentially the same as that obtained from CuO; i.e., they consisted of a single reduction process, but for given experimental conditions (β = 13 K min^{-1}, 10%v H_2/N_2) T_m was shifted to lower temperatures. The values of T_m recorded were 586, 583, 578, 556, and 566 K, respectively, for the series Cr, Mn, Fe, Co, and Ni, compared with 606 K obtained for CuO under the same conditions.

These first-row transition metals are all elements whose oxides have a more negative free energy of formation than CuO [35,36] and are therefore not expected to reduce before CuO [37]. For this reason their promoting influence is not thought to arise from either a hydrogen dissociation or a nucleation process involving the preferentially reduced dopant metal. Furthermore, the similarity of T_m observed for these samples is inconsistent with mechanisms involving the specific interaction of hydrogen with ionic dopant species. Thus, the results for H_2/D_2 exchange indicate high activity for Cr, Co, and Ni oxides with a minimum for Mn, Fe, and Cu oxides, while for hydrogen oxidation only Co_3O_4 is more active than copper oxide [38-41]. The effect of these ions would seem to arise from a nonspecific modification of the CuO lattice, resulting in more potential copper nucleation sites.

The TPR profiles obtained for the remaining doped oxides (10%v H_2/N_2, β = 13 K min^{-1}) are shown in Fig. 13. The Pd- and Ru-doped samples may be distinguished from the others in that reduction is complete at temperatures 100-200 K below the temperature of complete reduction found for CuO, and the TPR profiles consist of only two peaks (strongly overlapping for the Pd-doped sample), which indicates a low-temperature process

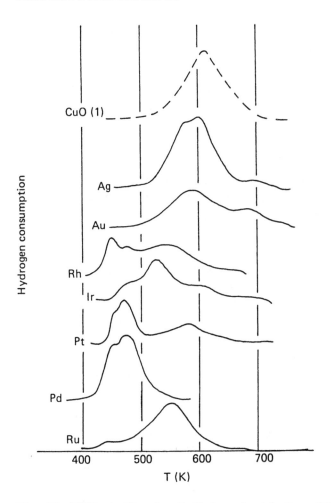

Fig. 13 TPR profiles for CuO doped with 2 mol% of the indicated metal. (From Ref. 14.)

followed by a more intense high-temperature reduction. For the remaining samples, Ag, Au, Rh, Ir, and Pt, the TPR profiles observed are more complex with reduction complete at about the same temperature as that observed for CuO. One member of each of the two groups was studied in detail, and for this purpose Pd and Pt were chosen.

1. Palladium-Doped Copper Oxide

The TPR profiles for Pd-doped samples are shown in Fig. 14. At low levels of Pd doping, 0.01 to 0.2% Pd, the TPR profiles appear as a major peak with a high-temperature shoulder. As the concentration of palladium is increased within this range, both the major peak and the shoulder move to lower temperatures. The promoting influence of palladium may arise in two ways. Palladium ions may distort the copper oxide lattice to produce potential copper nucleation sites such that reduction occurs at a lower temperature but does not involve the preferential reduction of palladium ions. However, the ease of reduction of bulk palladium oxide (reduction occurring at room temperature) suggests preferential reduction of palladium ions to form palladium nucleation sites.

The emergence of a new low-temperature process above 0.5% Pd is taken to support the mechanism involving the preferential reduction of palladium ions. The results indicate that Pd^{2+} does not reduce immediately on contact with H_2 at room temperature but is reduced to Pd^o simultaneously with the reduction of some Cu^{2+} to Cu^o in the low-temperature peak of the TPR profile. Thus, it is proposed that the promoting influence of palladium arises from the preferential reduction of Pd^{2+} (together with some Cu^{2+} reduction) leading to the production of palladium-rich nucleation sites.

2. Platinum-Doped Copper Oxides

The TPR profiles for Pt-doped samples are shown in Fig. 15. At low levels of Pt doping (0.01 and 0.05%) the TPR profiles are similar to those described for Pd. As the loading is increased a

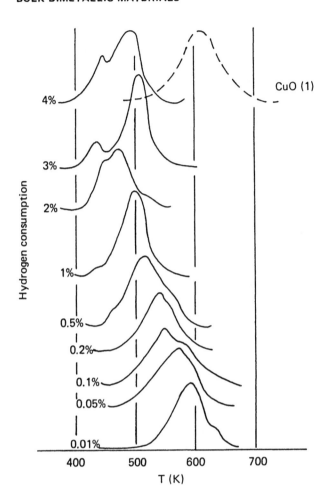

Fig. 14 TPR profiles for CuO doped with x mol% of Pd. (From Ref. 14.)

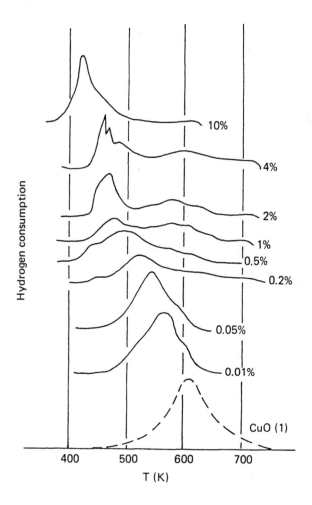

Fig. 15 TPR profiles for CuO doped with x mol% of Pt. (From Ref. 14.)

low-temperature peak emerges, again as observed for Pd; however, this peak grows to dominate the profile by 4% Pt. The emergence of a high-temperature tail at 0.2% Pt distinguished the TPR profiles of Pt from those of Pd-doped samples.

For Pt contents below 0.05% Pt the results are consistent with the mechanism proposed for palladium-doped samples. The development of the low-temperature peak indicates that Pt^{4+} is reduced to Pt^{o} simultaneously with some $Cu^{2+} \rightarrow Cu^{o}$ in this first process. However, powder X-ray results indicate a new phase in the 4% Pt-doped samples. It is therefore proposed that the copper oxide lattice is less able to tolerate Pt^{4+}, so that phase separation occurs, producing a platinum-rich phase and essentially pure copper oxide. The TPR profiles thus reflect the reduction of the Pt-rich phase followed by reduction of CuO modified to some extent by the presence of Cu-Pt alloy particles.

3. Other Group VIII Metal-Doped Copper Oxide

The results for CuO-2-Rh are similar to those obtained for the platinum-doped samples (Fig. 13). This is taken to indicate a similar phase separation into a rhodium-rich mixed oxide and essentially pure copper oxide.

The TPR profile for CuO-2-Ir also suggests the presence of two phases. However, powder X-ray experiments indicate the presence of IrO_2. This, coupled with the small low-temperature peak, suggests that the two phases are essentially IrO_2 and CuO.

The TPR profile for CuO-2-Ru is similar to the profiles obtained for the palladium-doped samples in that there is no evidence of a high temperature "tail." However, the appearance of faint lines in the powder X-ray results may indicate the presence of RuO_2 as a distinct phase. Furthermore, the small peak at 440 K is in agreement with published results for a ruthenium oxide Adams catalyst.

4. Group Ib Metal-Doped Copper Oxide

The TPR profiles for CuO-2-Ag and CuO-2-Au both show the characteristic high-temperature tail indicative of phase separation.

However, the major reduction processes do not commence until about 100° above the onset of reduction for the group VIII metal-doped samples. Within group Ib the free energies of formation of the oxides become less negative with increasing atomic number. Therefore, it would be expected that both silver and gold reduce preferentially to produce nucleation sites in a manner similar to the group VIII metal-doped oxides.

5. The Role of the Dopant Metal

A comparison of the results obtained for the group VIII- and group Ib-doped samples indicates that in both series the dopant ion is preferentially reduced to produce nucleation sites for the bulk reduction process. However, group VIII metals promote the reduction to a larger extent than do the group Ib metals. Within each group, alloys with copper are well known, for example, Cu_3Pt and Cu_3Au [42]. Thus, the distinction between the groups does not appear to arise from the stability of the reduced phase. It is concluded that the group VIII metals have a greater influence on the rate of growth of nuclei, once formed, than do the Ib metals.

A mechanism is proposed in which hydrogen is dissociatively adsorbed on the reduced group VIII metal islands of the dopant-rich phase, followed by spillover onto the oxide. In the case of Pd-doped samples, where no phase separation occurs, reduction proceeds exclusively by an intraparticular spillover mechanism. Consequently, the TPR profiles are simple with reduction complete at temperatures below that for pure CuO. For the other group VIII dopants, e.g., Pt, where phase separation occurs, it is proposed that the dopant-rich phase is reduced by a similar process, followed by less efficient interparticular spillover onto the essentially pure copper oxide phase. It is the reduction of this copper oxide phase that produces the characteristic high temperature "tail."

IV. SUPPORTED MONOMETALLIC MATERIALS

In the field of heterogeneous catalysis often the catalytic component is present in only very small quantities. Also, in the field of metal catalysis the metal component is usually required to be present in as high a dispersion as possible, that is, the maximum possible surface of the catalytic metal is exposed to the reactants. To achieve this situation catalytic components are usually dispersed on an "inert" support since this can lead to very high dispersions. The reason for qualifying the word *inert* is that in some cases the support has a more active role, e.g., in forming compounds with the catalytic component or the surface of the support being changed as a result of treatment to which the catalyst is subjected.

This quality of support catalysts—very high metal dispersions and often low concentrations of active catalytic component—can make it very difficult to characterize such catalysts by conventional methods. TPR, with its great sensitivity and the fact that it imposes no constraints on the catalyst other than that it be reducible, is an admirable technique for the characterization of such materials.

In this section we review the work done in this area using TPR. By and large all of the work has been concerned with the characterization of catalyst materials. However, in our opinion, the TPR technique could be applied with the same degree of success in the study of corrosion, for instance, which can be envisioned as reducible materials supported on a metal base.

The section is arranged like Section II; i.e., we will deal first with monometallic supported materials and second with bimetallic or doped materials.

A. Iron

The reduction of Fe_2O_3/SiO_2 has been studied by TPR [43]. The reduction profile consisted of two peaks at about 580 and 720 K which correspond to the processes

$$H_2 + 3Fe_2O_3 \rightarrow 2Fe_3O_4 + H_2O$$

and

$$4H_2 + Fe_3O_4 \rightarrow 3\,Fe + 4H_2O$$

In other words, in this catalyst which contained around 5%w Fe, the metal is completely reduced in a two-stage process.

A more recent study by Van't Blik [31] looked at the reduction of 3.5%w Fe on SiO_2. The TPR was done on the catalyst after the impregnation stage. The reduction started at 473 K and consisted of two peaks at 573 K and 638 K which were assigned to the reduction of $Fe(NO_3)_3$ for the low-temperature peak and Fe_2O_3 for the high-temperature peak, and the extent of reduction was only 42% of that expected for complete reduction to Fe metal. The reduced samples were then oxidized by heating in a programmed fashion to 773 K in a stream of O_2/He prior to carrying out a second TPR. In this case, which is comparable to the work reported in Ref. [43], reduction began at 493 K and by 773 K only 30% of the Fe^{3+} species was reduced. Van't Blik inferred that the reduction of well-dispersed Fe species on oxide supports is difficult owing to inhibition of the nucleation, i.e., the reaction between molecular hydrogen and the metal oxide to produce metal atoms and water.

There is some difficulty in reconciling the work of Unmuth et al. [43] with that of Van't Blik [31]. Although the concentrations of the metal were in the same region, and the supports very similar, the results found and conclusions reached are quite different. The only experimental difference between the respective works is that Unmuth et al. did TPR on dried, 723 K calcined catalysts whereas Van't Blik performed TPR directly on the impregnated samples followed by a programmed oxidation to 773 K. It could be that during the first TPR the exothermic effect of direct reduction of nitrate followed by the programmed oxidation to 773 K caused a strong interaction with the support, rendering subsequent reduction more difficult. McNicol [unpublished results] also found that bulk Fe_2O_3 physically mixed with γ-Al_2O_3 was also reduced only to the extent of about 50% up to temperatures of 1000 K.

The TPR of Fe supported on γ-Al_2O_3 has also been studied [44]. In this work the influence of calcination temperature and iron content on the solubility of Fe^{3+} in the Al_2O_3 lattice was studied. The main part of the Fe^{3+} species is dissolved at 473–593 K and the solubility of Fe^{3+} is constant at 593–938 K. This is the case irrespective of Fe^{3+} content at 665 K. Surface Fe_2O_3 species were generally different from the bulk Fe_2O_3, as evidenced by the quite different TPR spectra of supported and unsupported Fe-containing materials.

B. Nickel

1. Silica-Supported Nickel

Robertson et al. [9] were the first to report TPR results for Ni supported on SiO_2, in particular 0.25%w Ni on SiO_2. They found a stoichiometric reduction of Ni^{2+} to Ni° but at temperatures (peak maximum 680 K) much higher than that found for bulk NiO. They concluded that this indicated a strong interaction between Ni and the support. For higher loadings (5%w Ni on SiO_2) three TPR peaks were found [43] termed α, β, and γ. The α peak (535 K) was interpreted as arising from an endothermic phase transition from rhombohedral to cubic NiO occurring simultaneously with the reduction. The β peak (600 K) was attributed to the process:

$$H_2 + NiO \rightarrow H_2O + Ni$$

while the γ peak (663 K) was thought to arise from the reduction of very small oxide particles or possibly silicate species. Reoxidation of the reduced Ni catalyst produced a single peak at 553 K, indicating that the cubic phase of NiO is stabilized at room temperature by the reoxidation procedure. The absence of the γ peak in such samples was interpreted as indicating that metal support interactions were weakened by reoxidation.

The TPR of a standard 1%w Ni/SiO$_2$ catalyst was measured in four different laboratories as part of the work of the Society of Chemical Industry Working Party on Catalyst Reference Materials [45]. The profiles obtained by the four laboratories were all slightly different (Fig. 16), but the conditions used varied significantly from laboratory to laboratory in terms of heating rate, flow rates, and gas composition. In fact, the results found are a useful indication of the influence of experimental conditions on catalyst evaluation by TPR. It cannot be emphasized enough that for proper comparison of materials by TPR (as with any other technique) the conditions of measurement must, by and large, be comparable from sample to sample.

Perhaps the most detailed study to date of TPR applicability to the characterization of Ni catalysts has been the work of Mile and Zammitt [46] at Liverpool Polytechnic where a comprehensive series of experiments were conducted on Ni/SiO$_2$ catalysts. These workers showed that TPR was able to reveal characteristics of the nickel metal and nickel oxide phases that were not readily detected by other complicated and more expensive spectroscopic methods. It was concluded that the technique could prove to be a very useful quality control for use in the plant. In fact, they showed how TPR discriminated readily between active and inactive batches of cracking catalyst even when other techniques showed no perceptible differences. They thus used TPR extensively as a diagnostic tool for changes caused by variables during the preparation stages used for catalysts prior to actually reducing them. They were able to conclude from these studies that the TPR profiles were indicative of the actual activity and selectivity of the final reduced metal catalyst. Also, they drew a number of conclusions of importance to large-scale catalyst manufacture. It is, in our opinion, worth taking the space here to go into the detail of this work.

The TPR of a 9%w Ni on SiO$_2$ catalyst after calcination is shown in Fig. 17. There are four reduction peaks at, respectively, 503, 588, 658, and 823 K. The peaks were assigned as follows: The 503-K signal arose from Ni^{3+} reduction and its

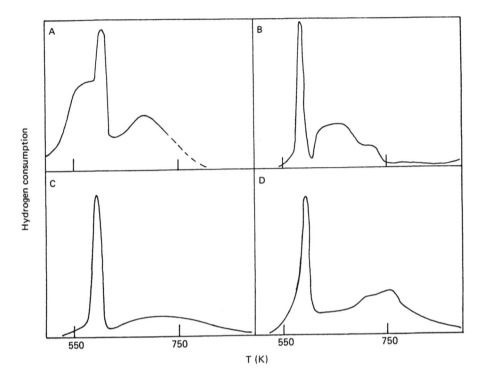

Fig. 16 TPR profiles for the uncalcined Ni/SiO$_2$ catalyst precursor. Experimental parameters (A) 5% H$_2$, 5 K min^{-1}, (B) 10% H$_2$, 30 K min^{-1}, (C) 25% H$_2$, 7 K min^{-1}, (D) 25% H$_2$, 27 K min^{-1}. (From Ref. 45.)

appearance was associated with black regions in the generally green NiO. It did not appear in all catalysts but its presence did not influence the position and proportion of the other peaks. The second peak, at 588-K, arises from bulk NiO reduction and is identical to a peak produced by reduction of calcined bulk Ni(NO$_3$)$_2$. Also, as this peak grows in intensity, X-ray powder diffraction revealed the presence of bulk NiO. The third peak, at 658 K, was usually the main peak at loadings of more than 1%w Ni and was assigned to NiO which had appreciable interaction with the SiO$_2$. The final peak, at 823 K, was assigned to a much stronger NiO interaction with the support.

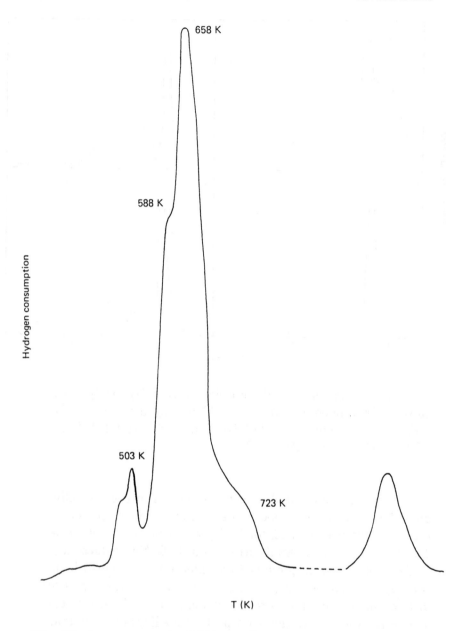

Fig. 17 A representative TPR profile of nickel on silica, 9.2%w, calcined at 593 K for 30 min. (From Ref. 46.)

Mile and Zammitt were able to show that the peak at 823 K was associated with NiO located in the larger pores (>100 μm) of the SiO_2. This is contrary to what one would expect if the ease of reduction were simply governed by the transport of H_2 and H_2O in pores of different dimensions of the SiO_2—in other words, the Ni species in the large pores would be easiest to reduce if this were the controlling factor. The authors were able to demonstrate this by comparing the TPR of Ni supported on a silica having large pores (diameter ~80 μm) and high pore volume (1.89 cm^3 g^{-1}) with that of a silica with small pores (diameter ~1.5 μm) and small pore volume, where it could be seen that the proportion of Ni involved in a strong interaction with the support was much greater in the silica with the large pores.

The amount of Ni in strong interaction with the silica increased with decreased loading and increased calcination temperature, which suggests that a chemical rather than a transport effect is dominant.

Regarding the actual preparation procedure for the catalysts, Mile and Zammitt showed that variations in the impregnation stage had very little effect on the TPR profile and activity of the catalyst. Likewise, preparation of catalysts via sequential impregnation produced similar performance in the TPR as catalysts prepared in one impregnation. It could therefore be concluded from this that change in impregnation routines has little impact on the properties of the catalyst and is not a major source of catalyst irreproducibility. As a consequence, perhaps this stage did not require too stringent a control in large-scale preparations. On the other hand, drying procedures were shown to be important at low loadings, especially in silicas with large pores where retexturing is easier. With more robust silicas and at high Ni loadings the drying procedure was less crucial. The conclusion reached was that in large-scale manufacture each silica must be considered individually and the drying stage procedure tailored for each case.

The influence of calcination temperature was also studied. It was shown that as calcination temperature increased the reduction became more difficult and metal particles of reduced species were smaller (Fig. 18). However, it became increasingly difficult to actually reduce all the inventory of Ni. It was therefore concluded that the matter was equivocal, but from the point of view of catalyst manufacture the lowest calcination temperature compatible with complete nitrate decomposition should be chosen. This would reduce the severity of the subsequent reduction and obtain a good dispersion of the metal.

At calcination temperatures above 873 K the presence of water vapor had an effect on the TPR profiles. The effect was to reduce the temperature of reduction and to increase the Ni particle size. Similar effects have been found in other catalyst systems. The water appears to reduce the formation of the surface nickel responsible for the high dispersion of the metal. The presence of water vapor during calcination could have crucial consequences in large-scale manufacture and could cause inhomogeneity and irreproducibility because of the difficulty of ensuring that the drying of a large batch of catalyst is uniform.

Increase in the metal loading led to an increased proportion of Ni arising from the low-temperature reduction peaks (Fig. 19) and as a consequence larger Ni particle sizes.

The catalytic and selectivity measurements of Mile and Zammitt were in general agreement with their conclusions drawn from the TPR work. Their work represents one of the first comprehensive applications of TPR to a single system.

2. Alumina-Supported Nickel

The reducibility of Ni on γ-Al_2O_3 is much more difficult than on silica. Mile and Zammitt [46] prepared some samples of Ni on a γ-Al_2O_3 from Ketjen but the catalysts did not show *any* reduction peaks up to 893 K (the maximum temperature their apparatus reached). The interaction of Ni with Al_2O_3 is clearly much stronger than the interactions that occur with silica

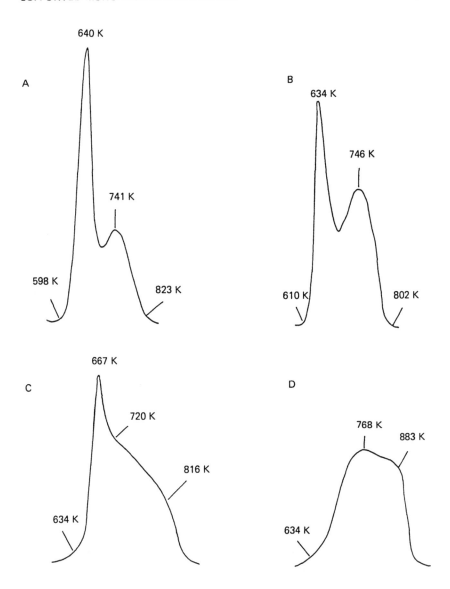

Fig. 18 Effects of calcination temperature on the TPR profile of 1% nickel on micronized Gasil 35 silica. (A) 623 K, (B) 673 K, (C) 773 K, (D) 973 K. (From Ref. 46.)

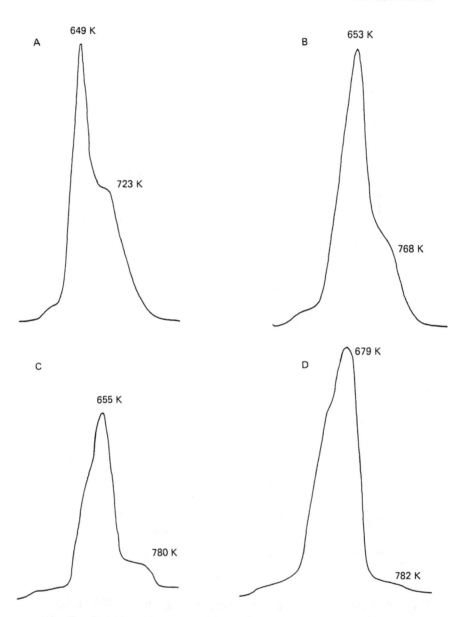

Fig. 19 Effects of nickel loading on TPR profiles. Micronized
Gasil 35. Standard conditions (A) 1%w, (B) 5%w, (C) 9.5%w,
(D) 31%w. Reduced sensitivity from A to D. (From Ref. 46.)

referred to above. Similar results were obtained for Al_2O_3-supported Ni by Paryjczak et al. [47].

C. Copper

1. Silica-Supported Copper

Silica-supported CuO has been studied using TPR [9], and it was found that for a 0.75%w Cu-loaded catalyst the position of the TPR maximum of the reduction peak was lowered by about 70 K to 570 K compared with that of bulk CuO, with the reduction profile somewhat broader. The support in this case appeared to function purely as a dispersing agent, enhancing the reactivity of the oxide in the reduction.

More recently, a study has been made [48] of Cu on silica prepared by ion exchange. Three TPR peaks were observed, one at 538-573 K, termed the α-peak; a second at 673 K, the β-peak; and a third at 923-1073 K, the γ-peak. The results were interpreted as arising from the reduction of Cu^{2+} interacting strongly with the silica surface in a stepwise fashion to Cu^o through Cu^+. This process resulted in the α- and γ-peaks. Isolated Cu^{2+} ions weakly interacting with the silica surface gave rise to α- and β-peaks.

If the catalysts were precalcined at more than 1073 K a fourth peak, called α^1, appeared at 603-623 K and α-, β-, and γ-peaks diminished in intensity. This was assigned to reduction of bulk CuO which crystallized from the species mentioned above, i.e., the reducibility of Cu on silica depended critically on the preparation method.

Delk and Vavere [49] studied catalysts of composition 1-4%w Cu on silica using TPR. The catalysts were calcined at 723 K prior to reductions. For 1%w Cu on silica the main reduction peak was found at 500 K with a shoulder at 550 K. The consumption of hydrogen was equivalent to the reduction of Cu^{2+} to Cu^o. The lower temperature found by these workers compared with that reported by Robertson et al. [9] can be explained by the greatly increased reducing gas flow rate used by

Delk and Vavere (95 ml min^{-1} compared with 10 ml min^{-1}).
Delk and Vavere [49] also studied 1-4%w Cu on titania. The
copper species reduces at a much lower temperature on the support than on silica. There is a reduction peak at 403 K in 1%
Cu on TiO$_2$ while in the 4%w sample there is a second peak at
493 K (Fig. 20). There is a shoulder appearing at this temperature in the 1%w Cu samples. The hydrogen consumption
is consistent with a reduction from Cu^{2+} to Cu^o. The low-
temperature peak was ascribed to reduction of Cu^{2+} species in
intimate contact with the titania support. The titania support
apparently catalyzes the Cu^{2+} reduction. The higher-temperature
peak arises from reduction of Cu^{2+} species that are not in such
close contact with the support. An interesting observation
found in this work was the further consumption of hydrogen
at temperatures above 573 K, i.e., after all the Cu is reduced.
This was interpreted as being due to the reduction of the titania
support catalyzed by copper metal. This reduction of the support leads to suppression of the activity of Cu on titania catalysts
for aldehyde hydrogenation.

D. Germanium

Germanium is an important constituent of modern bimetallic
platinum-containing reforming catalysts. There have been no
TPR studies of the unsupported oxides, but Ge supported on
γ-Al$_2$O$_3$ has been studied as part of a wider study of the reducibility of bimetallic reforming catalysts [50]. Catalysts containing 0.25% w Ge on γ-Al$_2$O$_3$ after drying and calcining at 798 K
were shown to reduce in a single peak at 873 K. The hydrogen
consumption was 40% less than that required for complete reduction of Ge^{4+} to Ge^o. This may imply that the Ge species is
involved in a strong interaction with the γ-Al$_2$O$_3$ support;
zeolite aluminium germanates have been synthesized, so apparently compound formation between Ge and Al species is not
difficult. However, in the absence of TPR data for bulk GeO$_2$,
it is premature to speculate on compound formation.

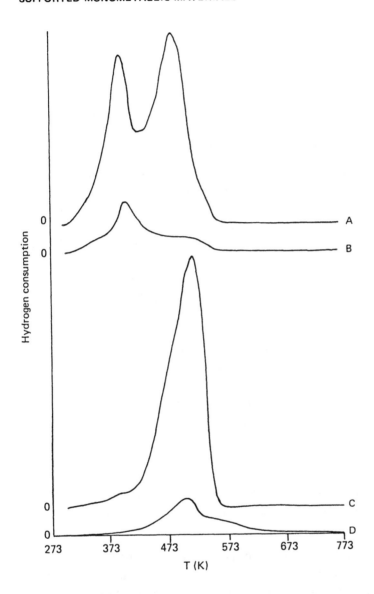

Fig. 20 TPR of supported copper catalysts following calcination at 723 K, (A) 4%w Cu/TiO$_2$, (B) 1%w Cu/TiO$_2$, (C) 4%w Cu/SiO$_2$, (D) 1%w Cu/SiO$_2$. (From Ref. 49.)

E. Ruthenium

Ruthenium, as mentioned earlier, is catalytically important for a number of reactions, for instance, in ammonia synthesis, Fischer-Tropsch hydrocarbon synthesis, chlorine production, and as a promoter of methanol electro-oxidation catalysts. Like most metal catalysts, its activity is usually enhanced by placing it in some form or another on a suitable support.

Koopman et al. [51] in a study of Ru on SiO_2 catalysts using TPR showed that catalyst prepared by direct reduction of $RuCl_3$ on SiO_2 upon oxidation at room temperature oxidized first on the surface, followed by bulk oxidation. This behavior differed substantially from that of catalysts prepared by calcination of $RuCl_3/SiO_2$ at high temperature, followed by reduction, then exposure to air at room temperature. Application of a high temperature reduction (<923 K) of $RuCl_3/SiO_2$ increased the active area of metal owing to removal of Cl from the Ru surface. Reduction of RuO_2/SiO_2 catalysts obtained by calcination of $RuCl_3/SiO_2$ in air produced a lower dispersion of Ru/SiO_2 than those prepared by direct reduction of $RuCl_3/SiO_2$. The above results explained the effect of activation procedure on activity of Ru/SiO_2 catalysts for benzene hydrogenation.

In the case of electrocatalysts the ruthenium component is usually supported on a conducting support such as carbon. Recently there has been some work done on the characterization of methanol electro-oxidation catalysts containing ruthenium [19]. During this study TPR results were obtained on ruthenium catalysts prepared by impregnating carbon-fiber papers with complex ruthenium salts.

When $RuNO(NO_3)_x$ is supported on carbon-fiber paper, the reduction profile is altered [19] in such a way that the reduction of the dried sample becomes more difficult, indicating an interaction between the Ru species and the carbon paper surface oxidized by the HNO_3 in the impregnating solution. When the sample is air-activated at 573 K, the TPR profile exhibits a series of peaks between 453 and 573 K corresponding to the reduction of the oxide and undecomposed nitrate (Fig. 21). This supported oxidelike species of Ru is thus more difficult to reduce than unsupported RuO_2, and it seems that the oxidelike ruthenium species is involved in an interaction with the oxidized surface of the carbon support. Blanchard et al. [52], in a recent study of the reducibility of H_2RuCl_6 impregnated on SiO_2 and γ-Al_2O_3, found that the Ru species was more difficult to reduce on γ-Al_2O_3 than on SiO_2, confirming that the interaction between metal salts and support is generally stronger with γ-Al_2O_3 than with SiO_2.

Bossi et al. [18] studied the reducibility of ruthenium supported on magnesium oxide (MgO). They claimed that their study showed the power of TPR in demonstrating the interaction between precursors, metals, and supports during the preparation of supported catalysts. They also pointed out the possible danger of doing TPR using H_2/N_2 gas mixtures when the material being reduced contains components that might be catalytically active for the synthesis of ammonia.

The above workers used $RuNO(NO_3)_x$ as the impregnating agent and prepared catalysts using a variety of procedures. It is not completely clear from their work exactly what the procedures were or how the TPR was actually carried out. However, we will attempt to interpret the results found. A catalyst prepared by impregnation followed by heating at 673 K under helium and then oxidized at 423 K in a helium/oxygen mixture produced a TPR profile with two peaks, at 473 K and 650 K, respectively. The hydrogen uptake corresponded to that expected for the reduction of Ru^{4+} to Ru^o. Upon successive reduction via TPR up to 1073 K followed by reoxidation at 423 K profiles were produced with reduction peaks at much lower temperatures (360 K) than the original sample. It seems that the repeated reduction-oxidation process leads to a weakening in the Ru-MgO interaction originally formed since the final reduction peak at 360 K is substantially lower in temperature than that of bulk RuO_2.

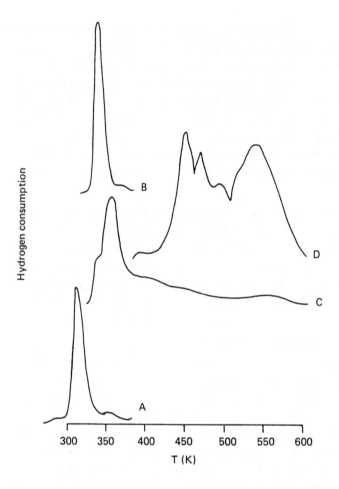

Fig. 21 TPR profiles for Pt/Ru/carbon-fiber paper catalysts
dried at 393 K and activated at 573 K. (A) 100%w Pt, (B)
87%w Pt, (C) 74%w Pt, (D) 100%w Ru. (From Ref. 19.)

F. Rhodium

Yao et al. [53] have studied rhodium supported on γ-Al_2O_3 catalysts using TPR. This was in support of an investigation of this system in vehicle exhaust catalysts. In such catalysts rhodium is used in small amounts as a selective component for NO reduction. Thus, interactions that may occur between rhodium and the γ-Al_2O_3 support at elevated temperature under oxidizing conditions are of interest. Yao et al. showed that when Rh/γ-Al_2O_3 catalysts are oxidized at temperatures above 873 K, some of the rhodium undergoes an interaction with the support so that the rhodium is buried in the support and becomes difficult to reduce, reducing in fact above 773 K under the conditions of the TPR experiment. Figure 22 shows the TPR profile of the 0.92% w Rh/γ-Al_2O_3 sample where the effect is most marked. In higher concentration samples, e.g., 5.51% w Rh/γ-Al_2O_3, a smaller fraction of the rhodium is involved in this interaction. In samples calcined below 773 K the reduction stoichiometry corresponded to a three-electron reduction. These results showed the limits of practical application of such catalysts under oxidizing conditions and indicated that during calcination compound formation between catalytic metal and support had taken place.

Very recently, Vis [33] completed a study of 2.3%w Rh supported on γ-Al_2O_3, TiO_2, SiO_2, and La_2O_3 using TPR. In the case of Al_2O_3-supported samples he studied the passivation phenomena of such catalysts and was able to comment on the degree of oxidation of the reduced catalysts after exposure to oxygen at room temperature. After TPR Vis subjected the catalysts to a programmed oxidation treatment up to high temperatures and obtained temperature-programmed oxidation profiles of the catalysts. After this treatment subsequent TPR runs showed that for the Al_2O_3-supported catalysts a single reduction peak was observed at 360 K with a hydrogen consumption equivalent to the reduction of Rh^{3+} to Rh^0. Unsupported Rh_2O_3 reduced in a single peak at 400 K. The result for Al_2O_3-supported Rh compares with the reduction peak at 373 K found by

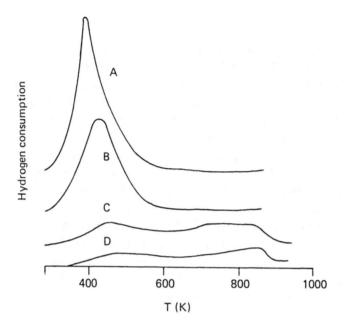

Fig. 22 TPR profiles of 0.92%w Rh/γ-Al₂O₃. Calcination treatment: (A) 673 K, 2 h, (B) 773 K, 16 h, (C) 873 K, 16 h, (D)
973 K, 16 h. (From Ref. 53.)

Yao et al. [53]. These latter workers used a heating rate of 8 K
min^{-1} compared with 5 K min^{-1} used by Vis.

   In the case of 3%w Rh on SiO_2 catalysts the oxidized
catalyst showed a single peak at 335 K attributed to the reduction of well-dispersed Rh_2O_3. For TiO_2-supported catalysts two
reduction peaks were observed at 325 and 385 K. The low-
temperature peak corresponds to that found for SiO_2-supported
catalysts and arises from well-dispersed Rh_2O_3 while the high-
temperature peak was attributed to reduction of bulk Rh_2O_3.
It was also apparent from the TPR results that some reduction
of the TiO_2 support takes place during the TPR run. It is significant that the dispersion of Rh in the reduced catalysts was
much poorer for Rh/TiO_2 than for Rh/Al_2O_3 and Rh/SiO_2,
which is in accord with the TPR results discussed above.

The passivation studies showed that for the well-dispersed Rh/Al_2O_3 catalysts, passivation is rather drastic (Fig. 23) such that long passivation times (like storage in air) comes close to real oxidation. For TiO_2-supported catalysts, which have much greater Rh crystallite sizes, passivation results only in the formation of an outer layer of oxide on top of relatively large metal particles.

For 2%w Rh on La_2O_3 catalysts the TPR peak occurred at much higher temperatures than for any other system studied by Vis [33]. A single peak occurred at 560 K with a hydrogen consumption greater than that expected for Rh_2O_3. It was speculated reasonably that a strong interaction occurred between Rh_2O_3 and La_2O_3 during oxidation to produce a species more difficult to reduce than bulk Rh_2O_3. The higher consumption of hydrogen could be attributed to some reduction of the La_2O_3 support.

In a detailed study [33] of Rh/Al_2O_3 and Rh/TiO_2 catalysts of varying metal loading ranging from low to high, TPR was able to demonstrate how the reduction properties of the metal could be related to the dispersion of the metal. This was particularly so for the TiO_2-supported catalysts (Fig. 24). The appearance of the second reduction peak at 385-400 K is clearly associated with increasing amounts of bulk Rh_2O_3 in such catalysts.

G. Rhenium

There has been some controversy in the literature concerning the reducibility of rhenium in supported catalysts, in particular regarding the extent of its reduction [54-57]. Bolivar et al. [57], Webb [55], and McNicol [54] have claimed that rhenium supported on γ-Al_2O_3 is completely reduced to the zero valent state by hydrogen at temperatures ranging from 673 to 823 K. Johnson and Le Roy [56] reported that rhenium on γ-Al_2O_3 is reduced by hydrogen to the Re^{4+} state exclusively. It has become apparent that the discrepancies arise from the different conditions of preparation and reduction of the catalysts involved. TPR has

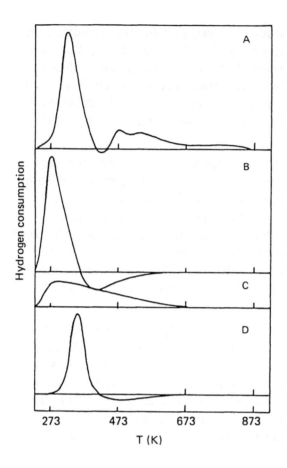

Fig. 23 2.3%w Rh/Al$_2$O$_3$ catalyst. (A) TPR of passivated catalyst prereduced at 473 K. (B) TPR of passivated catalyst prereduced at 773 K. (C) TPO following TPR of the catalyst.
(D) TPR following TPO of the catalyst. (From Ref. 33.)

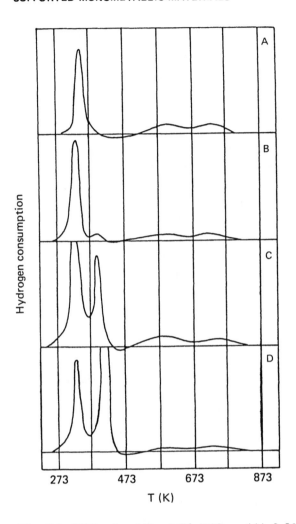

Fig. 24 TPR of oxidized Rh/TiO₂. (A) 0.3%w, (B) 1.0%w, (C) 3.2%w, (D) 8.1%w. (From Ref. 33.)

proved to be a very useful tool in identifying these influences on the condition of Re in typical supported catalysts.

Yao and Shelef [25] examined the reducibility of ReO_3 and 1.21%w Re/γ-Al_2O_3. They found that ReO_3 reduces in the range 623–648 K (Fig. 25) with an uptake of hydrogen corresponding to a six-electron reduction. Similarly, Re/γ-Al_2O_3 catalysts prepared by impregnation and drying at 573 K showed the same TPR profile but with a high-temperature tail (Fig. 25). If the catalyst was subjected to a dispersing treatment (773 K oxidation in air followed by a nitrogen purge), then a different TPR profile was obtained (Fig. 25). This time the rhenium component was reduced in a peak at about 823 K. After dispersion, rhenium is involved in a strong interaction with the γ-Al_2O_3, making its reduction more difficult. Bolivar et al. [57] found that Re/γ-Al_2O_3 catalysts that had only been dried at 393 K produced similar results to those obtained from the undispersed catalyst of Yao and Shelef [25], i.e., reduction of the rhenium component at low temperatures. McNicol [54], in a study of 0.2%w Re/γ-Al_2O_3, showed that if the catalyst was prepared by impregnation, drying, and calcining at 798 K in air followed by a pure nitrogen purge or drying, the rhenium component reduced in a peak at 823 K with a hydrogen consumption equivalent to a seven-electron reduction, which is analogous to the dispersed catalyst of Yao and Shelef [25]. Again it seems that the role of the calcination step is to fix the rhenium component to the γ-Al_2O_3 surface in some kind of compound formation such that upon subsequent reduction at high temperatures a good dispersion of rhenium is obtained. Similar interactions with rhenium and SiO_2 do not appear to occur [58] since all the rhenium in such catalysts can be reduced below 773 K.

Arnoldy et al. [59] have recently demonstrated using TPR the strong metal-support interaction that occurs between Re and Al_2O_3. In this work the reduction peak for the Al_2O_3-supported metal appeared at higher temperatures than for either SiO_2- or carbon-supported Re, indicating the greater strength of the metal-support interaction in the case of Al_2O_3. The reduction

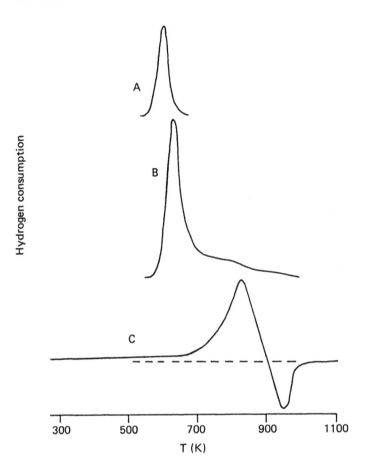

Fig. 25 TPR profiles of Re catalysts. (A) Unsupported ReO_3,
(B) 1.21%w Re/Al_2O_3, (C) 1.21%w Re/Al_2O_3 before dispersion,
(D) 1.21%w Re/Al_2O_3, after dispersion at 773 K in flowing
nitrogen for 16 h. (From Ref. 25.)

temperature found for well-dispersed Re on Al_2O_3 in this work was 730 K, substantially lower than in the earlier work discussed above. This is explainable in terms of the much higher hydrogen concentrations used by Arnoldy et al. in their TPR measurements. These authors proposed too that the varying reducibilities of Re on Al_2O_3 reported [54–57] were due to the presence of additives such as chloride, which was thought to increase the Re^{7+}-support interaction.

In a later publication Arnoldy et al. [60] measured activation energies for reduction of various Re on Al_2O_3 catalysts using TPR. They were able to demonstrate good agreement with values calculated via other techniques. Activation energies found for reduction of surface ions in Re_2O_7/Al_2O_3 catalysts were higher than for NH_4ReO_4 crystallites on the surface, in agreement with the strong metal-support interaction in the former. However, the increase in activation energy and its variation with Re content were not as substantial as expected, given the apparently strong interaction with the support. This was put down to the influence of initially reduced Re in autocatalyzing the further reduction of the metal.

The activation entropy for reduction of Re_2O_7/Al_2O_3 catalysts increases with increasing Re content [60]. This was explained by assuming that the freedom of rotation of the activated complex (Re-O-H) was restricted by the surface geometry, particularly at low Re content. The TPR results were related to the metathesis activity of Re_2O_7/Al_2O_3 catalysts as a function of Re loading.

H. Molybdenum, Chromium, and Tungsten

In a study of MoO_3 on SiO_2 Thomas et al. [61] showed that the TPR peak occurred at 750 K as opposed to 900 K for bulk MoO_3. However, on going to increasingly higher MoO_3 loadings a peak appeared in the TPR of the supported catalysts, which could be ascribed to reduction of bulk MoO_3. Work with γ-Al_2O_3-based catalysts reduced at higher temperatures than SiO_2-based catalysts [62] confirmed the observations made for many other systems that stronger interactions occur between metals and γ-Al_2O_3 than between metals and SiO_2. The results indicated that at low loadings on SiO_2 the support served to disperse the metal on the surface, but as loading increased this dispersion deteriorated.

In the case of WO_3 on SiO_2 TPR showed that the tungsten species reduced at 900-950 K compared with a temperature of 800 K found for bulk WO_3 [61].

For both systems these workers correlated the reducibility [defined as the temperature under TPR conditions at which half of the amount of hydrogen required for complete reduction of Mo^{4+} (W^{6+}) to Mo^0 (W^0) is consumed] with the hydrodesulfurization (h.d.s.) activity.

The easier the reducibility of the catalyst, the higher the h.d.s. activity. This is connected with the influence of good catalyst dispersion on the activity of the sulfided phase. A similar relationship holds for the butene hydrogenation activity.

Van Roosemalen et al. [63] also studied the reducibility of 4-6%w WO_3 on SiO_2 catalysts using TPR. The catalysts were prepared by three different routes, namely (1) impregnation, (2) reaction between WCl_6 and SiO_2, and (3) cogelation of WCl_6 and tetraethoxysilane with HCl. These catalysts are active for alkene metathesis. In cases (1) and (2) peaks were found at 680-780 K and 865 K. The low-temperature peak was ascribed to metathesis inactive oxide species while the high-temperature peak was ascribed to 12-silicotungstic acid, a surface compound that is probably the precursor to the active site for metathesis. Case (3) was distinctly different in that only a very small fraction (<5%) of inactive oxide was present. A high-temperature peak at 1050 K was ascribed to a surface compound between the WO_3 and SiO_2 that was highly active for metathesis. Similarly, using TPR and laser-Raman spectroscopy Thomas et al. [64] also concluded that the precursor for the active site in cases (1) and (2) was a surface compound and not the free oxide.

In a study of Cr/SiO_2 ethanol oxidation catalysts Parlitz et al. [65] found two TPR peaks, one at about 633 K and the other at 753 K. The high-temperature peak was ascribed to reduction of chromate species stabilized on the surface, whereas the low-temperature peak found at increasing Cr concentrations arose from reduction of di- or polychromate species. Comparisons with catalytic activity pointed to the chromate species as the active site for the reaction.

A study of MoO_3 and WO_3 on γ-Al_2O_3 [62] using TPR showed that the reducibility of MoO_3 catalysts prepared by impregnation was related to surface coverage of the support. This implies either that interaction between the Mo species and the Al_2O_3 is greatest at lower coverages or that the effect is caused by differences in the degree of aggregation, which increases with surface coverage. It was not possible to differentiate between the two possibilities.

Reduction of WO_3 on γ-Al_2O_3 was less easy than that of MoO_3 species on Al_2O_3 and this probably reflects a stronger interaction between W and the support [62].

I. Cobalt

Paryjczak et al. have studied the TPR of 1–10%w Co_3O_4 on SiO_2 and Al_2O_3 [30]. For 10%w Co_3O_4/SiO_2 the reduction peak is at 693 K and reduction is complete by 873 K. For the 1%w sample reduction is still incomplete at 873 K. This reduction behavior compares with that for bulk Co_3O_4, which reduces in a double peak with maxima at 593 and 663 K, respectively. Clearly the increased difficulty of reduction in the supported samples reflects an interaction between the Co species and the SiO_2 support. For γ-Al_2O_3-supported samples reduction was much more difficult. In the case of 10%w Co_3O_4, reduction was still incomplete at 923 K. This is yet another example of the much stronger interaction that tends to occur between Al_2O_3 supports and metal species.

A more recent study of 4%w Co on Al_2O_3, 2%w Co on TiO_2, 4%w Co on SiO_2 was carried out by Van't Blik [31].

Reduction of oxidized Co on Al_2O_3 was difficult starting at
473 K and still not complete at 773 K. The TPR profile is
characterized by two peaks, one at 573 K and the other at 673
K. Comparison with the profile for bulk Co_3O_4 showed that for
the supported catalyst the reduction is also a sequential process.
In order to reduce the catalyst completely the temperature had
to be raised to 1373 K and hydrogen was consumed between
1073 and 1373 K, probably arising from the reduction of a surface Co_3O_4 spinel. In a study of reduced catalysts that had
been passivated in air, the TPR results clearly indicated that during passivation corrosive chemisorption of oxygen had taken
place. A similar observation was made regarding passivation of
TiO_2-supported catalysts. The oxidized Co/TiO_2 catalyst reduced
in two peaks, at 588 K and 686 K, that were ascribed to reduction of a fraction of Co_3O_4 (588 K) to CoO and reduction of
CoO (686 K) to Co^0. This explanation was confirmed as being
correct by stopping the TPR at 640 K (the minimum between
the two peaks), cooling to 223 K, then running a second TPR
when only the 686 K peak was observed. The results of this
series of tests are shown in Fig. 26. If during the first TPR
metallic Co had been formed, then in the second run a shift of
the higher-temperature peak to lower temperatures would have
been expected since any metallic Co species would have acted
as nucleation centers for the dissociation of hydrogen molecules,
thus facilitating the reduction. No such shift was observed (Fig.
26), confirming Van't Blik's explanation.

Recently Arnoldy and Moulijn [66] reported on Co/Al_2O_3
catalysts and their conclusions were broadly similar to Van't
Blik. They suggested that four phases existed in such catalysts:
Phase I reducing around 600 K was ascribed to Co_3O_4 crystallites; phase II at about 750 K consists of Co^{3+} in crystallites of
stoichiometry Co_3AlO_6 or in a well-dispersed surface species;
phase III reducing at about 900 K arises from surface Co^{2+} ions;
and phase IV reducing around 1150 K from either Co^{2+} in a
diluted Co-Al spinel or in $CoAl_2O_4$.

In the TPR of oxidized Co/SiO_2 catalysts the reduction
starts at 473 K and is complete at 973 K. The higher

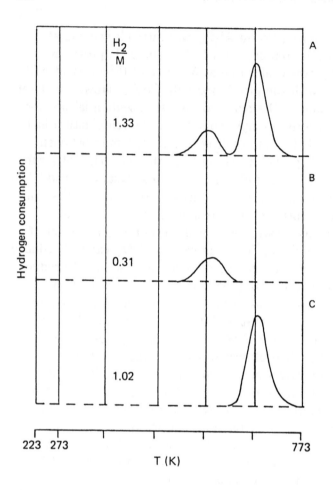

Fig. 26 TPR profiles of Co(2.01%w)/TiO$_2$ catalyst. (A) TPR of oxidized catalyst. (B) TPR until 640 K. (C) TPR of the catalyst that was partly reduced up to 640 K. (From Ref. 31.)

temperatures needed for reduction of the supported catalyst compared with bulk Co_3O_4 is in accordance with a metal-support interaction—a similar conclusion to that of Paryjczak et al. [30].

J. Vanadium

Vanadium oxides are widely used in catalysts for oxidation reactions, e.g., oxidation of sulfur dioxide, carbon monoxide, and hydrocarbons. Usually the vanadium component is supported on an oxidic support.

Roozeboom et al. [27] have made a detailed study of the TPR of V_2O_5 supported on a wide variety of supports such as γ-Al_2O_3, SiO_2, and TiO_2 using a range of catalyst preparation methods. On Al_2O_3 the TPR shows a single peak at low vanadium concentrations (<1.7%w) at about 650 K. As the concentration rises above this level a second peak appears at 600 K. Above 3.7%w a third peak appears beyond 700 K. In conjunction with Raman spectroscopy studies it was established that the 650-K peak arises from a two-dimensional octahedral polyvanadate surface structure. The peak above 700 K was assigned to crystalline V_2O_5. The fact that this peak occurs below that of bulk V_2O_5 was ascribed to the smaller crystalline size of the supported species. Tilley and Hyde [67] showed that the rate of reduction of V_2O_5 increases with decreasing crystallite size. The vanadium content of the catalysts calculated from TPR, assuming reduction from V^{5+} to V^{3+}, agrees excellently with X-ray fluorescence analysis of vanadium content.

On SiO_2, likewise, TPR peaks are found that can be assigned to surface phase and crystalline V_2O_5 species, respectively. A similar situation exists on TiO_2 supports.

Roozeboom et al. [27] also studied catalysts prepared by an ion exchange method. For SiO_2 again both a dispersed surface phase and crystalline V_2O_5 were found. On Al_2O_3 the catalysts prepared by ion exchange showed only the two surface vanadium species giving rise to the peaks at 600 and 650 K. There was no peak at 700 K for crystalline V_2O_5. This points

to a complete monolayer coverage with tetrahedral and poly-
meric octahedral vanadate phases. For TiO_2 the TPR profile
was identical with that of the impregnated catalysts.

The main conclusions drawn from this work is that TPR
was able to give a quantitative analysis of the crystalline V_2O_5
and the respective vanadium surface phases whereas Raman spectroscopy enabled only a qualitative analysis to be made.

K. Silver

TPR has been used to study 4%w Ag on γ-Al_2O_3 [68]. Three
peaks were observed in calcined samples. A double peak at
370–500 K was ascribed to different oxide species of Ag while
the peak at 870 K arose from Ag species involved in strong
metal-support interaction with the Al_2O_3.

L. Platinum

1. Silica-Supported Platinum

In the Pt/SiO_2 system, catalysts can be made by impregnation
or ion exchange with the surface hydroxyl groups of the SiO_2.
Usually the catalysts prepared by ion exchange lead to better
dispersions of platinum upon subsequent reduction, as might be
expected because in ion exchange there is a true chemical interaction between the platinum species and the support. Jenkins
et al. [69] examined this system in detail. Their results for the
ion-exchanged and impregnated Pt/SiO_2 catalysts are shown in
Fig. 27. The impregnated catalyst was made using H_2PtCl_6
while the ion-exchanged catalyst was made using [Pt(NH_3)_4]
(OH)_2 aqueous solution. The TPR profiles for the two catalysts
are quite different. The ion-exchanged catalyst reduces over a
higher temperature range (Fig. 27)—higher even than unsupported
[Pt(NH_3)_4](OH)_2—and displays structure. It seems from the
relatively high temperature of reduction that the platinum
species in this catalyst is interacting with the SiO_2 support. The
hydrogen consumption corresponded to a two-electron reduction.
The impregnated catalyst reduced in three widely different

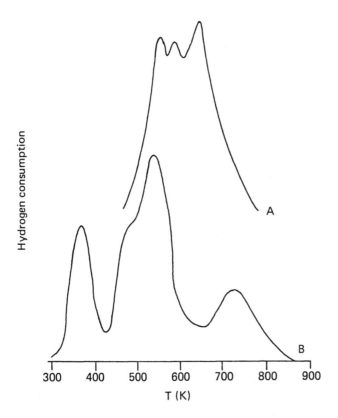

Fig. 27 TPR profiles of Pt/SiO$_2$ catalysts. (A) Ion-exchanged
Pt/SiO$_2$, (B) impregnated Pt/SiO$_2$. (From Ref. 69.)

temperature regions (Fig. 27) but with a hydrogen consumption
corresponding to a four-electron reduction. The low-temperature
reduction peak corresponds to that found with unsupported bulk
PtCl$_4$ and H$_2$[PtCl$_6$], and upon reduction this would produce
fairly massive platinum particles that would not contribute significantly to the metal dispersion. Thus, the poorer dispersion
of the metal in the impregnated catalyst can be ascribed to bulk
H$_2$[PtCl$_6$] or PtCl$_4$ still present after the drying step. Jenkins
et al. [69] were also able to show, using TPR, how by appropriate calcination procedures the bulk PtCl$_4$ or H$_2$[PtCl$_6$] in the

impregnated catalyst could be made to interact with the SiO_2 to produce a well-dispersed catalyst, thus demonstrating the necessity for a calcination step in the preparation of such catalysts.

More recently Foger and Jaeger [70] studied the reduction of 0.9–2.8%w Pt on SiO_2 catalysts. The particle sizes of the metal in the reduced catalysts were observable by X-ray diffraction placing them at >3 nm diameter. The TPR of reduced catalysts exposed to reoxidation at temperatures 373 K < T < 973 K showed no reduction peaks. This was put down to the ease of reducibility of such large reoxidized Pt particles, i.e., they reduced below room temperature and were thus inaccessible to the TPR method used by Foger and Jaeger.

2. Alumina-Supported Platinum

Platinum supported on γ-Al_2O_3 is of considerable importance, particularly in the well-established catalytic reforming process for improving the octane rating of gasoline. In this application catalysts containing <1%w Pt are employed. The catalysts are usually prepared by impregnation of γ-Al_2O_3 with H_2PtCl_6, drying in air at 393 K, followed by calcination at 798 K. This carefully set-out procedure results in the production of a catalyst that, after reduction with hydrogen, contains platinum particles that are not detectable by transmission electron microscopy. It is thought that this excellent dispersion of platinum is achieved because of an interaction between γ-Al_2O_3 and the platinum species that occurs during the calcination step [see, for example, Ref. 71], but the nature of this interaction is not well established. McNicol [54], Blanchard et al. [52], and Yao et al. [32] have studied the TPR of this system.

McNicol [54] showed that catalysts containing 0.375%w Pt on γ-Al_2O_3 after drying and calcining reduced in a single peak at 553 K with a hydrogen consumption corresponding to a four-electron process. He concluded that since this temperature was so far above the reduction temperature of the unsupported tetravalent platinum chlorides and oxides, there must be an interaction between the platinum species and the γ-Al_2O_3 surface, and that this interaction took place during the calcination step. However, McNicol did not examine the TPR profile of catalysts that had not been calcined.

Recently Blanchard et al. [52] studied the TPR of 2%w Pt/γ-Al_2O_3 prepared from $H_2[PtCl_6]$ by impregnation. For calcined catalysts they obtained TPR profiles similar to those of McNicol, i.e., with a main reduction peak at about 553 K. Moreover, in the samples that had been dried at 383 K but not calcined, a similar profile was also obtained. This result would imply that the interaction between the Pt species and the γ-Al_2O_3 surface took place during the drying step. However, this contradicts evidence from other studies that showed that the development of the best dispersion of platinum depended on a certain minimum time of calcination at a temperature of about 783 K [72]. Indeed, J. W. Jenkins [unpublished results] has examined the TPR of a solid prepared by mixing together Al $(NO_3)_3$ and $H_2[PtCl_6]$ and calcining at 773 K in air and found a profile very similar to that of calcined Pt/γ-Al_2O_3. It seems clear, therefore, that an interaction takes place between the platinum species and the γ-Al_2O_3 support that fixes platinum chemically to the support so that upon reduction the Pt^0 species remains in intimate contact with the γ-Al_2O_3.

Yao et al. [32] explored this issue using TPR in conjunction with other techniques. In low-loading calcined samples they found that the Pt species reduced in a single peak at around 523 K in reasonable agreement with McNicol [54] and Blanchard et al. [52]. This contrasted with the facile reduction at around room temperature of their unsupported PtO_2 catalysts. In high-loading catalysts they observed reduction peaks at much lower temperature, i.e., 293 K and 363 K. They argued that the low-load, highly dispersed phase reducing at 523 K arose from the reduction of a $PtO_2 \cdot Al_2O_3$ complex.

Huizinga et al. [72] studied 5%w Pt on Al_2O_3 prepared via a combined ion exchange and impregnation technique using

compounds that did not contain Cl. In dried catalysts they
found evidence using TPR for a number of Pt species with a
net valency greater than 2 but less than 4. The main reduction
took place at 473 K. As the temperature of calcination increases
the TPR peak moves to lower temperature and the oxidation
state of Pt to higher values. It was shown that the PtO_2 particles
interact with the Al_2O_3 such that the PtO_2 does not decompose
after calcination at 773 K. This conclusion was confirmed by
temperature-programmed oxidation studies on a reduced Pt/Al_2O_3
catalyst that showed that the PtO_2 formed did not start to decompose until temperatures greater than 1023 K. This is about
100 K higher than the decomposition point of pure PtO_2. Thus,
again the interaction between Al_2O_3 and Pt was demonstrated
and in this case for catalysts prepared in the absence of Cl-containing materials. The actual ease of reduction of the calcined
samples in this study is greater than in other studies using much
lower concentrations of Pt and usually an impregnation technique
based on H_2PtCl_6. While the same general conclusions can be
drawn about interactions with Al_2O_3, it is clear that the TPR
technique enables the fine differences between catalysts prepared
by different methods to be established.

Lieske et al. [73] studied Pt on Al_2O_3 catalysts of loading
0.5-1.0%w in which some catalysts contained Cl and others did
not. Calcination of Cl-containing catalysts at 773 K gave TPR
profiles remarkably close to those found by McNicol [54] and
Yao et al. [32] for similarly prepared samples. Lieske et al. [73]
proposed that the TPR peak was due to reduction of a Pt/O/Cl-
containing surface complex stabilized by interaction with Al_2O_3.
This was also supported by UV-Visible spectroscopic studies.
The authors also showed via studies on Cl-free catalysts that reoxidation of a Cl-free sample at 473-573 K produced a TPR
profile similar to that of a Cl-containing sample, thus indicating
that the species produced on reoxidation of the reduced catalyst
is Cl-free. However, reoxidation at 773-823 K produced a surface species of PtO_2. A treatment with HCl followed by drying
and oxidation at 723 K gave the same TPR profile as a freshly

calcined Cl-containing catalyst. Their work showed in fact that a Cl-redispersal step is essential in regeneration of Pt/Al_2O_3 reforming catalysts.

3. Carbon-Supported Platinum

The combination of platinum with a carbon support has been studied using TPR [74]. The catalysts were prepared by two methods, one involving the impregnation of a carbon-fiber paper support with $[Pt(NH_3)_4](OH)_2$ solution followed by drying at 393 K and calcination at 573 K, the other involving ion exchange of a preoxidized carbon-fiber paper support with aqueous solutions of $[Pt(NH_3)_4](OH)_2$ followed by washing, drying at 393 K, and calcination at 573 K. The catalysts that had been only dried at 393 K displayed TPR profiles that showed reduction of the platinum species at temperatures slightly in excess (for the impregnated catalysts) and markedly in excess (for the ion-exchanged catalyst) of that of bulk $[Pt(NH_3)_4](OH)_2$ (Fig. 28). In the ion-exchanged catalyst the interaction between the $[Pt(NH_3)_4]^{2+}$ cation and the acidic surface oxides on the oxidized carbon paper renders reduction more difficult, i.e., analogous to $[Pt(NH_3)_4]^{2+}$ exchanged onto SiO_2.

The impregnated catalyst, while showing a TPR peak at slightly higher temperature than bulk $[Pt(NH_3)_4](OH)_2$, also shows a weaker peak at the same temperature as bulk $[Pt(NH_3)_4](OH)_2$, indicating the presence of some bulk $[Pt(NH_3)_4](OH)_2$ in the sample.

After calcination in air at 573 K, the reduction characteristics were of particular interest in that subambient reduction peaks were obtained. For impregnated catalysts (Fig. 29), reduction occurred in two peaks at 177 and 195 K with a hydrogen consumption corresponding to less than a two-electron reduction. The ion-exchanged catalyst (Fig. 29) showed a single peak at 217 K with a hydrogen consumption corresponding to a two-electron reduction. The shift to higher temperature of the ion-exchanged catalyst can again be ascribed to the stronger

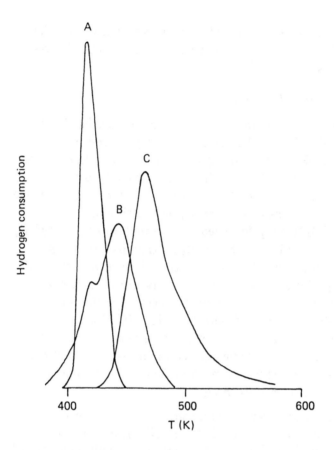

Fig. 28 TPR profiles of Pt/carbon-fiber paper catalysts dried at 393 K. No activation. (A) Unsupported $Pt(NH_3)_4(OH)_2$. (B) Carbon-fiber paper impregnated with $Pt(NH_3)_4(OH)_2$. (C) Carbon-fiber paper ion-exchanged with $Pt(NH_3)_4(OH)_2$. (From Ref. 74.)

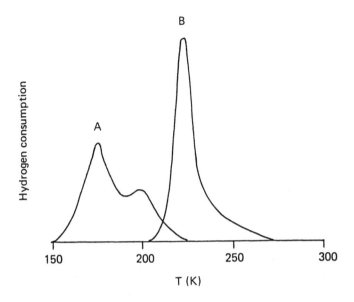

Fig. 29 TPR profiles of Pt carbon-fiber paper catalysts dried at 393 K and activated at 573 K. (A) Carbon-fiber paper impregnated with $Pt(NH_3)_4(OH)_2$, (B) carbon-fiber paper ion-exchanged with $Pt(NH_3)_4(OH)_2$. (From Ref. 74.)

interaction between the Pt species and the carbon support. After reduction, the ion-exchanged catalyst contained platinum crystallites of mean size 3.0 nm while the impregnated catalyst had a mean platinum size of >10 nm. Again, the stronger interaction between platinum and support, as shown by TPR, leads to better metal dispersion, analogous to the Pt/SiO_2 system. Unsupported $[Pt(NH_3)_4](OH)_2$ when activated at 573 K in air leads to the formation of platinum metal as shown by X-ray powder diffraction and the absence of reduction peaks in TPR. Clearly the lower stoichiometry of reduction for the air-activated impregnated catalyst arises because some platinum metal already exists at this stage, presumably formed from decomposition of the small amount of bulk $[Pt(NH_3)_4](OH)_2$ identified in the TPR of the dried catalyst.

Unsupported $Pt(NH_3)_2(NO_2)_2$ when air activated also leads to platinum metal species, but when supported on carbon paper and air activated produces Pt^{4+} species that reduce above room temperature. Clearly the decomposition of both supported salts in air leads to the formation of quite different platinum species. The differing reducibility characteristics described above must, in our opinion, have significant implication for the design of suitable preparation, activation, and reduction conditions for such supported catalysts.

4. Tin Oxide-Supported Platinum

Information on the influence of the calcination step in catalyst preparation was provided by a TPR study of the Pt/SnO_2 catalyst system by Hughes and McNicol [24]. It was found that a high-temperature calcination (573–773 K) in air rendered the Pt species difficult to reduce, so that treatment with H_2 at 573 K or electrochemical reduction at room temperature did not reduce the Pt species completely. As a result, the catalysts had almost zero activity for the electro-oxidation of methanol, the reaction studied by Hughes and McNicol. The TPR profiles obtained for the dried and calcined H_2PtCl_6/SnO_2 catalysts, together with the profile for pure SnO_2, are shown in Fig. 30. It can be seen that the Pt species in the dried sample reduces in a temperature region slightly higher than that for unsupported $PtCl_4$ and H_2PtCl_6 (Fig. 10), while for the 773-K calcined sample only weak reduction peaks attributable to Pt reduction can be discerned. It seems that the Pt species reduces in the same region as SnO_2. The reduction peak for SnO_2 in Pt-containing samples occurs at a lower temperature than for pure SnO_2, indicating some catalysis of the SnO_2 reduction. The results can be interpreted in terms of an interaction between the Pt species and the SnO_2, brought about by the calcination procedure. The TPR results were confirmed by the electrochemical technique of cyclic voltammetry. Thus the TPR technique in this case was able to demonstrate why calcination was detrimental to the quality of the catalyst. Hurst [unpublished results] found similar results for 1%w Pt on SnO_2 support.

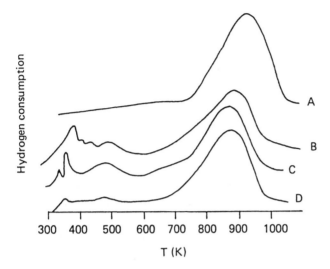

Fig. 30 TPR profiles for Pt/SnO$_2$ catalysts. (A) Pure SnO$_2$,
(B) 10%w Pt/SnO$_2$ (H$_2$PtCl$_6$ impregnated; dried at 393 K),
(C) dried Pt/SnO$_2$ catalyst, calcined at 573 K, (D) dried Pt/
SnO$_2$ catalyst, calcined at 773 K. (From Ref. 24.)

5. Titanium Oxide-Supported Platinum

Huizinga et al. [72] found for Pt on TiO$_2$ prepared by ion exchange and impregnation techniques that excessive hydrogen was
consumed in reduction of dried catalysts. Their conclusion was
that the support was undergoing reduction catalyzed by the Pt.
This confirmed earlier work using electron spin resonance [75],
which showed the presence of Ti^{3+} in Pt/TiO$_2$ catalysts reduced
below 500 K. Oxidized (673 K) samples of Pt/TiO$_2$ upon TPR
study indicates the presence of metallic Pt particles formed as a
result of the oxidation. This was confirmed by temperature-
programmed oxidation studies that showed that PtO$_2$ on TiO$_2$
decomposes at much lower temperatures than PtO$_2$ on Al$_2$O$_3$.
Oxidation at 673 K led to the formation of platinum metal particles with an outer oxide skin and a metallic core. This was
also the case when the catalysts were passivated where TPR
showed the presence of the metal core and outer platinum oxide
skin.

V. SUPPORTED BIMETALLIC MATERIALS

Bi- and multimetallic catalysts are of increasing importance owing to the promoting action that second and third components can have on the activity of a particularly active metal, usually by virtue of either alloy formation or dilution of the surface of the catalytically active metall The field of *supported* bimetallic catalysts is of prime importance since this is the main area of interest in real catalyst systems.

It is therefore vital to characterize the state of metallic components in such catalysts to determine the role of the second metallic component and also to establish what factors influence the relationship between the metals during catalyst preparation. Furthermore, the regeneration of spent bimetallic catalysts is more complicated than that of monometallics, as the interaction between the metals must be restored during the regeneration procedure. TPR has proved useful in characterizing catalysts where the active components are present as finely dispersed particles and in very low concentrations. There have been a number of investigations of supported bimetallic catalyst systems and the results are discussed in this section.

Often catalysts are modified by *small* amounts of promoters that are not usually termed bimetallic catalysts, perhaps because the promoter primarily modifies the support rather than the metal. Nevertheless, such promoters may influence the reducibility of the catalyst.

The overwhelming amount of work on supported bimetallic catalysts has been carried out on systems in which one of the metallic components is platinum. The bulk of this section will therefore be devoted to such systems.

A. Platinum-Based Bimetallics

1. *Platinum/Ruthenium*

TPR studies of supported Pt/Ru catalysts have been few. Blanchard et al. [52] examined the TPR of γ-Al$_2$O$_3$- and SiO$_2$-supported

Pt/Ru catalysts containing about 10%w metals. Dried and calcined catalysts were examined and Cl-containing salts were used
in the preparations. The TPR results of SiO_2-supported catalysts
showed that the ruthenium reducibility characteristics were hardly
altered compared with those of monometallic catalysts but that
the platinum reduction was more difficult. This was attributed
to the cocrystallization of platinum and ruthenium chlorides during the drying step of the preparation. For the γ-Al_2O_3-supported
catalysts the platinum reduction was unaffected by the presence
of ruthenium but the ruthenium reduction shifted to higher temperature, reducing in the same region as platinum. The relatively
high reduction temperature of the platinum species on γ-Al_2O_3
compared with that of ruthenium probably reflects the stronger
interaction between platinum and the γ-Al_2O_3 substrate.

Mahmood et al. [19] have studied Pt/Ru supported on
carbon-fiber paper, which is an active catalyst for methanol
electro-oxidation. In catalysts dried at 393 K it was shown that
the ruthenium reduction was catalyzed by the presence of platinum (Fig. 31). The broad reduction peak between 523 and
573 K for pure ruthenium on carbon disappeared in the bimetallic catalyst and the lower temperature peak was shifted. However, in dried-only catalysts the situation is complicated by the
nature of the salts used in the impregnation (nitrate, nitrite, and
nitroso). Thus, the air-activated catalysts where such groups
have been decomposed and the metals exist as oxides are more
informative of the nature of the catalyst. Pure Ru/carbon-fiber
paper activated in air at 573 K reduces in a series of peaks between 453 and 573 K, while pure platinum on carbon paper reduces at 313 K. In the bimetallic catalyst the reduction of the
Pt/Ru species takes place between 323 and 373 K, indicating the
influence of each component on the other's reducibility. The
platinum appears to catalyze the reduction of the ruthenium
component. The TPR profiles for these catalysts are shown in
Fig. 21. The finished catalyst (after air activation at 573 K and
electrochemical reduction) shows activity that is about 100 times
greater than that of the monometallic Pt catalyst. It also displays

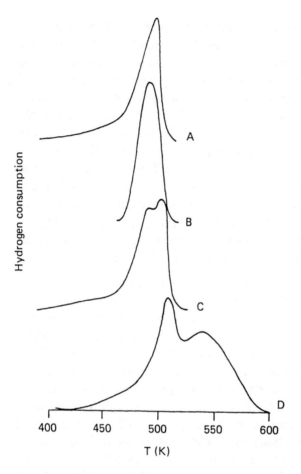

Fig. 31 TPR profiles for Pt/Ru/carbon-fiber paper catalysts
dried at 393 K. (A) 100%w Pt, (B) 85%w Pt, (C) 69%w Pt,
(D) 100%w Ru. (From Ref. 19.)

surface chemisorption, adsorption, oxidation, and reduction characteristics in cyclic voltammetry indicative of an alloyed Pt/Ru phase. Thus, the TPR evidence for alloying is independently confirmed.

2. Platinum/Rhenium

Platinum/rhenium bimetallic catalysts supported on γ-Al$_2$O$_3$ are used in modern catalytic reforming processes where they display better activity, selectivity, and stability than do the monometallic platinum catalysts. The enhanced performance is thought to arise from alloying between platinum and rhenium, although in such catalysts the metal species are present in such small amounts (<1%w) and are so finely dispersed that identification of alloying is not straightforward. It has also been questioned in the literature whether rhenium is completely reduced in such catalysts [54-57].

TPR studies of the Pt/Re on γ-Al$_2$O$_3$ system have been carried out by Bolivar et al. [57] and McNicol [54]. Bolivar et al. claimed that platinum catalyzed the reduction of rhenium in impregnated, dried catalysts so that the rhenium component could be reduced at relatively low temperatures (<573 K). McNicol, on the other hand, claimed that there was no catalysis of rhenium reduction in catalysts that had been dried and calcined at 798 K. The TPR profiles of 0.375%w Pt/0.2%w Re, 0.375%w Pt, and 0.2%w Re on γ-Al$_2$O$_3$ are shown in Fig. 32. Two peaks were observed in the TPR profile of the bimetallic catalyst, one at 553 K superimposing on the platinum reduction observed in the TPR profile of the monometallic platinum catalyst, and the other at 823 K superimposing on the rhenium reduction observed in the TPR profile of the monometallic rhenium catalyst. McNicol therefore concluded that there was no catalysis of the Re reduction by Pt. However, the reduction stoichiometry corresponded to the reduction of Pt^{4+} and Re^{7+} to the zero valent state and it was concluded that if alloying occurred, it must take place during the reduction step of the preparation.

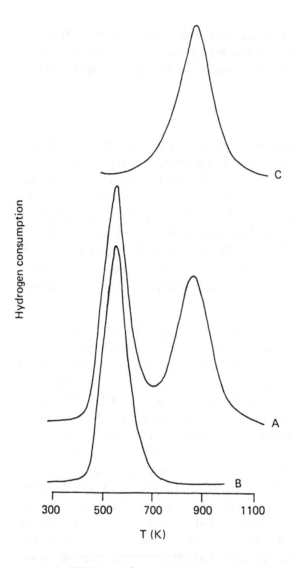

Fig. 32 TPR profiles of Pt, Re, and Pt/Re on γ-Al_2O_3 catalysts calcined at 798 K. (A) 0.375%w Pt 0.2%w Re/γ-Al_2O_3, (B) 0.375%w Pt/γ-Al_2O_3, (C) 0.2%w Re/γ-Al_2O_3.

Bolivar's finding that the rhenium species reduces at lower temperatures and is catalyzed by platinum has been ascribed to the increased degree of hydration of the oxidized metal species in catalysts that have been dried only at 393 K and not calcined. Support for this view has come from the same group of workers [76] who, following the procedure of McNicol—namely calcining at 798 K and purging with nitrogen prior to reduction—found that rhenium reduced exclusively above 773 K with no catalysis of the reduction by platinum.

Similar conclusions were reached by Wagstaff and Prins in their study of Pt/Re on γ-Al$_2$O$_3$ [77]. For an 0.35%w Pt/0.6%w Re on γ-Al$_2$O$_3$ catalyst dried in situ at 453 K, complete reduction was observed for both the Pt and the Re species around 523 K. In situ drying at 773 K, however, produced a TPR profile in which the platinum reduced around 523 K but the rhenium reduced at higher temperatures, around 773 K, confirming the results of McNicol. These authors further studied the effects of reoxidation of the catalysts. Treatment of the reduced catalysts with oxygen above 473 K causes segregation of platinum and rhenium. A subsequent TPR showed reduction of a platinum-rich species around 273 K and reduction of a rhenium-rich species around 523 K. Treatment with oxygen at 373 K, however, did not destroy the intimate contact between the platinum and rhenium. A subsequent TPR showed a single broad peak between 273 and 373 K. The implication of these results is that the reduction step is of critical importance in the preparation and regeneration of a Pt/Re reforming catalyst since the interaction between the metals apparently occurs during this step.

Isaacs and Peterson [78] using TPR showed also how the reduction of Pt/Re on Al$_2$O$_3$ catalysts was dependent on drying temperature. They interpreted their results as meaning that water influenced the rate of Re$_2$O$_7$ migration to Pt centers; i.e., the mobility of Re$_2$O$_7$ and hence its rate of diffusion to Pt centers was dependent on drying temperature. Even at drying temperatures up to 773 K a small amount of Re$_2$O$_7$ remained at least partially hydrated.

3. Platinum/Iridium

Platinum/iridium supported on γ-Al_2O_3 catalysts has been studied by Wagstaff and Prins [79]. The TPR profile of calcined 0.375%w Pt/0.37%w Ir on γ-Al_2O_3 showed a single reduction process around 508 K. This was expected since both calcined Pt/Al_2O_3 and calcined Ir/Al_2O_3 reduce in this region. Upon reoxidation of the reduced Pt/Ir catalyst and subsequent TPR a single reduction process was observed around 378 K, which was taken as evidence for the existence of bimetallic alloy clusters in the reduced catalyst.

Foger and Jaeger [70] disputed the above conclusions in that they claimed that only Ir oxides contributed to the TPR profiles between 298 and 623 K. According to them, only Ir was oxidized when Pt/Ir-supported alloys are heated in oxygen. Below 573 K surface oxides of IrO_3 are formed but above this temperature numerous individual crystals of IrO_2 are formed. Their conclusions were disputed by Wagstaff and Prins [79], who pointed out that the systems studied by Foger and Jaeger were predominantly SiO_2-supported rather than Al_2O_3-supported and also contained metal particles detectable by X-ray powder diffraction, i.e., mean size >3nm, whereas in the work of Wagstaff and Prins the metal particles were considerably smaller.

A further complication in this area is introduced in the work of Faro et al. [80] on Pt/Ir on Al_2O_3 catalysts. These authors used similar metal loadings and similar calcination conditions to those of Wagstaff and Prins, but they used Cl-free salts in the catalyst preparation. They found that after reduction and reoxidation below 523 K a single-phase TPR peak was found (Fig. 33), results analogous to those of Wagstaff and Prins. However, if reoxidation at 773 K took place, segregation of IrO_2 occurred. By and large their results are in agreement with those of Wagstaff and Prins, but Faro et al. disagreed with them on the issue of the valency of the Ir species being reduced. However, the catalyst preparations of Faro et al., as pointed out above, differed in that a Cl-free salt was used in their work. Nonetheless, the work on Pt/Ir catalysts and previous work by McNicol [54]

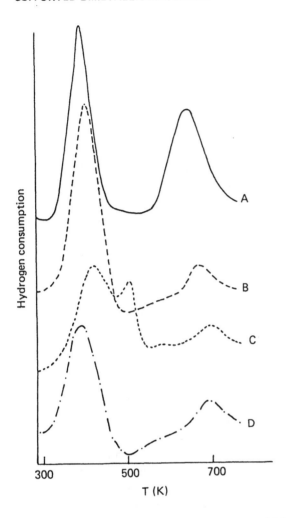

Fig. 33 TPR profiles of catalyst Pt/Ir (39%w Ir) under different pretreatment conditions. (A) Purging with 5% hydrogen/oxygen at room temperature, (B) calcination in 5% oxygen/helium at 523 K, (C) calcination at 773 K, (D) sample from C after re-oxidation at 523 K. (From Ref. 80.)

and Wagstaff and Prins [77] on Pt/Re-supported catalysts has shown that a calcination step at high temperature *prior to* reduction is crucial in producing good dispersions of the metal catalyst, particularly on Al_2O_3; yet a calcination at similar temperatures *after* reduction can have a deleterious effect on the interaction between the metals in the same catalyst.

As an aside to the above, Faro et al. in their work observed a peak at around 650 K (Fig. 33), which they attributed to the reductive elimination of hydroxy groups on the Al_2O_3 support. Such a peak has not been observed by others and may arise from the fact that no Cl-containing salts were used in the preparation of Faro and coauthors' catalysts.

4. Platinum/Germanium

Platinum/germanium catalysts supported on γ-Al_2O_3 are used in catalytic reforming [see, for example, Ref. 81] for reasons similar to those for using the Pt/Re catalysts. Again the question arises as to the exact nature of the interaction between the metals in such catalysts. TPR studies of this system have been carried out [50] with results similar to those for the Pt/Re systems. The TPR profile of 0.375%w Pt/0.25%w Ge on γ-Al_2O_3 calcined catalyst is shown in Fig. 34. There is a peak at 553 K corresponding to the reduction of the platinum component, and the hydrogen consumption corresponds to a four-electron reduction. A second peak at 873 K superimposes on the peak for reduction of Ge/γ-Al_2O_3 catalysts, and the hydrogen consumption is 40% less than that required for the reduction of Ge^{4+} to Ge^0. Thus, some oxidized germanium remains in combination with the γ-Al_2O_3 support. There is no catalysis of the germanium reduction by platinum, a finding similar to that in the Pt/Re system. However, if the catalyst after reduction is reoxidized at 453 K, TPR shows that the germanium component now reduces in the same region as the reoxidized platinum component. It would therefore seem that interaction or alloying between platinum and germanium occurs during the reduction step of the preparation.

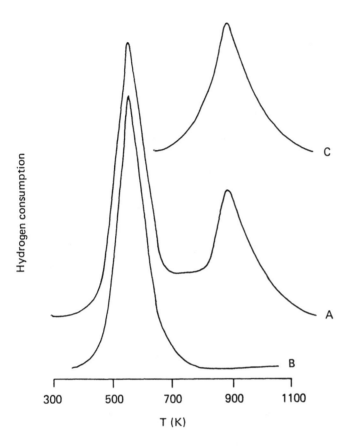

Fig. 34 TPR profiles of Pt, Ge, and Pt/Ge on γ-Al$_2$O$_3$ catalysts. (A) 0.375%w Pt, 0.25%w Ge/γ-Al$_2$O$_3$, (B) 0.375%w Pt/γ-Al$_2$O$_3$, (C) 0.25%w Ge/γ-Al$_2$O$_3$.

Support for this conclusion comes from hydrogen chemisorption measurements [82] that show that the H/Pt ratio for a
Pt/Ge catalyst has a value of 1 if the reduction temperature of
the calcined catalyst is 723 K but progressively falls as the reduction temperature rises into the region where the germanium component reduces. This would suggest that, as it is reduced, the
germanium component builds into the platinum surface, diluting
it and causing hydrogen chemisorption to fall. Thus, it would
seem that part of the germanium is completely reduced to Ge^0,
with the remainder interacting with the support as Ge^{4+}. Whether
the latter interaction is critical for the preparation of a successful
catalyst is not known.

It is clear that in bimetallic Pt/Re and Pt/Ge on γ-Al_2O_3-
supported catalysts the TPR profile is a useful indicator of catalyst condition in the calcined state. This is the condition to
which the catalyst should be returned when it is redispersed
after the carbon burn-off of a regeneration. TPR, then, could
be applied as a fairly rapid, reliable, and cheap indicator of the
success or failure of a particular regeneration process.

5. Platinum/Tin

Dautzenberg et al. [83] studied the platinum/tin system by looking at a range of supported catalysts containing 0.4 to 3%w total
metal. The catalysts were first reduced at 773 K then reoxidized
mildly at 353 K prior to TPR measurement. The profile for Pt/
Sn/Al_2O_3 differs markedly from that of either Pt/Al_2O_3 or Sn/
Al_2O_3 (Fig. 35) in that the reduction of the bimetallic occurred
intermediate between the two monometallics—a result similar to
that found by McNicol [50] for $Pt/Ge/Al_2O_3$ and Wagstaff and
Prins [79] for $Pt/Ir/Al_2O_3$ catalysts that had been reduced and
reoxidized. If on the other hand the catalysts were subjected to
severe reoxidation, then a separate SnO_2 phase was formed and
the TPR profile resembled that of the separate Pt/Al_2O_3 and
Sn/Al_2O_3 catalysts.

In the mildly reoxidized $Pt/Sn/Al_2O_3$ catalysts it is apparent
from the TPR profile that only a small amount of Sn is present

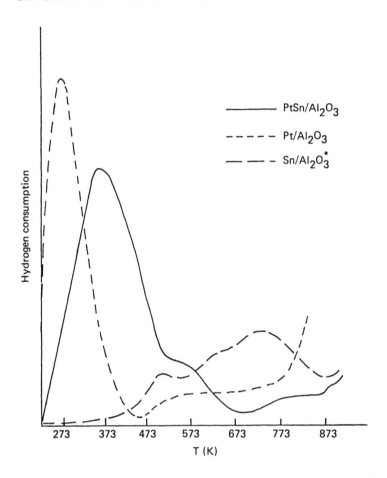

Fig. 35 TPR profiles after oxidation at 453 K of Pt, Sn, and Pt/Sn on γ-Al_2O_3 catalysts. ——— PtSn/Al_2O_3. — — — Pt/Al_2O_3. —— —— Sn/Al_2O_3 (it is estimated that about 30–40%w of the metal is reduced to zero valent Sn, assuming SnO_2 to be present upon oxidation). (From Ref. 83.)

as free SnO_2. The profile of the $Pt/Sn/Al_2O_3$ catalyst, mildly reoxidized, can be interpreted as arising from the reduction of Pt and Sn species in intimate contact with each other.

Burch also worked on the $Pt/Sn/Al_2O_3$ system [84] and arrived at totally different conclusions based on TPR work. It should be pointed out that Burch's procedure for preparing his catalysts was totally different from that of Dautzenberg et al. in that Burch added the Sn species to commercially prepared catalysts already containing Pt. In pure SnO_2/Al_2O_3 catalysts Burch showed that the Sn species could be reduced only to the extent of about 50%. Pure SnO_2 is known to reduce completely [76] and so Burch concluded that the Sn species was stabilized by interaction with the Al_2O_3. In the $Pt/Sn/Al_2O_3$ catalysts the Pt simply had the function of facilitating the Sn reduction but did not promote the complete reduction of the Sn species—this only occurring to the extent 50%. In contrast to the conclusions of Dautzenberg et al., Burch claimed that the Sn species was reduced only to the 2+ state and that the interaction between Sn^{2+} and Pt was responsible for the increased hydrogen chemisorption on Pt. He ascribed the special properties of Pt/Sn catalysts to the change in electronic properties of small Pt crystallites either by interaction of Pt with Sn^{2+} to give electron-deficient Pt or by incorporation of a small percentage of the Sn inventory into the Pt to give electron-rich Pt.

Burch and Dautzenberg et al. based their conclusions not solely on TPR evidence but also on information obtained from other techniques. For this reason, the differences in results cannot be ascribed to any obvious differences in the way in which the TPR technique was used. The problem appears to lie with the varying methods of catalyst preparation used by the different workers and also the totally different catalyst pretreatments used prior to the TPR runs. For example, Dautzenberg et al. [83] examined mildly reoxidized catalysts, whereas Burch [84] examined freshly calcined catalysts. The inescapable conclusion here is that it is quite wrong to draw conclusions about the correctness or incorrectness of the results of individual workers or the interpretations of results obtained by different groups when the work in the individual groups is not carried out in the same way. Such conclusions only serve to discredit the ability of techniques such as TPR in solving problems associated with the performance of catalysts. Where TPR has been carried out in different laboratories on materials or catalysts prepared or pre-treated in similar fashion, the reproducibility of TPR results be-tween laboratories is highly impressive and there are numerous examples of this cited in this book.

B. Rhodium-Based Bimetallics

1. Rhodium/Cobalt

Van't Blik has recently made a study of this system supported on Al_2O_3, SiO_2, and TiO_2 [31]. On Al_2O_3 after impregnation, the metal salts (~5%w total metal) as evidenced by TPR are in intimate contact (Fig. 36). Reduction produces bimetallic Co-Rh particles the outer skin of which is enriched in cobalt. Also, a part of the Co interacts with the Al_2O_3 to produce $CoAl_2O_4$, which cannot be reduced below 773 K. When the reduced cata-lyst is exposed to oxygen, cobalt is oxidized to a greater ex-tent and the resulting cobalt oxide covers a rhodium-rich core. In this situation relatively high temperatures are needed to re-duce the cobalt oxide because the covered Rh can no longer serve as a reduction catalyst. A thorough oxidation of the cata-lyst up to 773 K induces a remodeling of the bimetallic particles such that Rh_2O_3 is exposed and subsequent reduction is en-hanced (Fig. 36). During oxidation no additional $CoAl_2O_4$ is formed. After reduction of the oxidized catalyst bimetallic particles were formed.

In the case of TiO_2-supported Co/Rh catalysts (~3%w total metal), the TPR shows that the two metals reduce in the range 350–520 K after impregnation. This is the same tempera-ture region as that for the Rh catalyst. It was concluded that both metal salts are in intimate contact after this stage of cata-lyst preparation. After oxidation at 773 K Van't Blik found en-hancement of the Co reduction by Rh—the TPR profile was

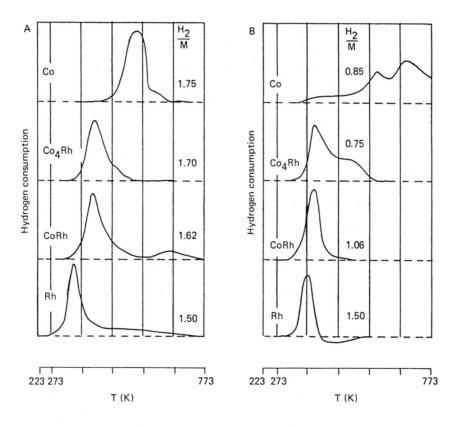

Fig. 36 TPR profiles of γ-Al$_2$O$_3$-supported Co (4.44%w), Co$_4$Rh (4.57%w), CoRh (5.4%w), and Rh (2.3%w) catalysts. (A) TPR of impregnated and dried catalysts, (B) TPR of oxidized catalysts. (From Ref. 31.)

similar to that of Rh on TiO_2. He concluded that the partly oxidized particles after this treatment consisted of either CoO + Rh_2O_3 or Co_3O_4 + Rh_2O_3 + Rh, in both cases with Rh ions on the surface. After 973-K oxidation the profiles resembled that of Co/TiO_2; that is, the bimetallic particles were covered with cobalt oxide during thorough oxidation.

For SiO_2-supported impregnated catalysts reduction occurred between 273 and 510 K. Again the rate of reduction of $Co(NO_3)_2$ was enhanced by Rh, suggesting that the Rh and Co salts were in intimate contact after impregnation. After oxidation the TPR showed a main peak at around 400 K with a shoulder at lower temperature. The reduction of the bimetallic was complete before the oxidized monometallic Co/SiO_2 started to reduce. Thus, the metals are close together after oxidation. In contrast with Al_2O_3-supported catalysts no irreducible compounds of Co with SiO_2 were formed.

Complementary Extended X-Ray Absorption Fine Structure (EXAFS) studies largely supported the above results and conclusions.

2. Rhodium/Iron

Rh-Fe on SiO_2 catalysts (~7%w total metal) were also studied by Van't Blik [31]. In the impregnated catalysts, reduction takes place between 273 and 473 K with the Rh catalyzing the reduction of $Fe(NO_3)_3$ and Fe_2O_3. After oxidation up to 773 K the catalyst showed a single TPR peak at 418 K with a degree of reduction of about 95%. The Rh species catalyzed the reduction of the Fe species, implying intimate contact between the two. Mossbauer spectroscopy confirmed that the Fe was present as Fe^0 in an Fe/Rh alloy.

Normally the reduction of well-dispersed Fe on oxide supports is difficult owing to the inhibition of the nucleation, i.e., the reaction between molecular hydrogen and the metal oxide to produce metal atoms and water. Consequently, degree of reduction of Fe on SiO_2 is low. The catalytic effect of the Rh on the Fe reduction is simply Rh atoms acting as nucleation centers.

3. Rhodium/Silver

Paryjczak et al. [68] studied the reducibility of Rh/Ag on Al_2O_3 catalysts (4%w total metal) using TPR. The profiles of the bimetallic catalyst were simply the sum of those of the individual monometallic components and they concluded therefore that there was no strong interaction between these two metals. According to Anderson et al. [85] Rh and Ag do not form alloys with each other.

C. Palladium-Based Bimetallics

1. Palladium/Nickel

Paryjczak et al. [86] investigated Pd/Ni on Al_2O_3 catalysts (total metal content 5%w) by TPR. The Pd species had a catalytic influence on the Ni reducibility as evidenced by the shift of the Ni TPR peak to lower temperatures in the bimetallic catalyst. It was therefore concluded that the two metal species interact with each other to form alloys.

2. Palladium/Lead

Karski and Paryjczak [87] have studied the Pd/Pb on Al_2O_3 system (10%w total metal) using TPR. Oxidation and thermal decomposition of bimetallic Pd/Pb/Al_2O_3 in a stream of argon at 773 K leads to segregation of individual oxides. After isothermal reduction at 823 K and reoxidation, TPR points to a bimetallic interaction between Pd and Pb atoms but only for small percentages of Pd.

D. Nickel-Based Bimetallics

1. Nickel/Copper

The determination of the reducibility and identification of alloying in copper-nickel-on-silica catalysts by TPR has been reported by Robertson et al. [9]. For a 0.75%w Cu/0.25%w Ni on silica catalyst, a pretreatment of 773 K for 1 h in air, followed by 1/2 h in N_2 at 673 K prior to cooling in nitrogen (treatment 1), produced a TPR profile in which all the Cu^{2+} and all the Ni^{2+} reduced separately (Fig. 37). However, treatment 1 for a 0.25%w Cu/0.75%w Ni sample resulted in a profile in which all the Cu^{2+} and Ni^{2+} reduced simultaneously (Fig. 37). This effect was thought to arise from the increased miscibility of CuO and NiO at lower Cu loadings.

Under other conditions, the 0.75%w Cu/0.25%w Ni catalyst produced TPR profiles in which the Cu^{2+} and Ni^{2+} reduced simultaneously. Thus, a sample that had been reduced following treatment 1 was further reduced for 16 h at 773 K then calcined at 673 K for 1/2 h (treatment 2). This produced a TPR profile in which all the Cu^{2+} and Ni^{2+} reduced simultaneously at around 473 K; a fresh sample was calcined externally at 773 K for 1 hr and cooled in air (treatment 3). The subsequent TPR profile simultaneously showed reduction of all the Cu^{2+} and Ni^{2+} at around 523 K. These effects were thought to arise from the hydration of the oxides on cooling in the air.

For samples with the same loading, but only dried at 373 K for 2 h, subsequent TPR produced simultaneous reduction of Cu^{2+} and Ni^{2+} as nitrates, with a consequent large increase in the hydrogen consumed. The lack of a reduction peak at higher temperatures (650–800 K) suggests that direct reduction of the dried copper and nickel nitrates gives rise to the formation of alloys. It was concluded that in all the copper-nickel-on-silica catalysts studied, both metals exist entirely in the zero valent state and are engaged in alloying on or after reduction at 773 K.

The Cu/Ni on γ-Al_2O_3 system has been studied recently by Paryjczak et al. [47]. The catalysts were calcined at 873 K followed by purging in argon at that temperature prior to cooling to room temperature and TPR measurement. The authors found that the Cu species had no significant effect on the reduction of the Ni. The very strong interaction found between the Ni and the Al_2O_3 support limited the potentially strong catalytic effect of Cu on the Ni species.

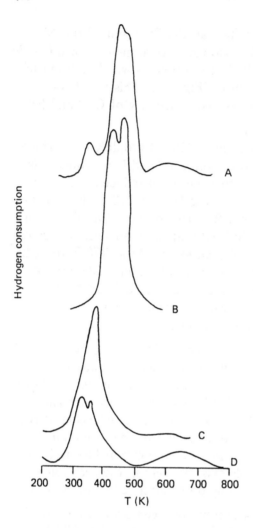

Fig. 37 TPR profile for Cu/Ni on silica catalysts. (A) 0.25%w Cu/0.75%w Ni after heating in air at 773 K, 1 h, followed by nitrogen at 673 K, 0.5 h, and cooling in nitrogen. (B) 0.75%w Cu/0.25%w Ni after heating in air at 773 K and cooling in air. (C) 0.75%w Cu/0.25%w Ni after treatment as in A, reduction for 16 h at 773 K, and calcination at 673 K, 0.5 h. (D) 0.75%w Cu/0.25%w Ni as for A. (From Ref. 9.)

2. Nickel/Iron

In a study of Fischer-Tropsch catalysts, Unmuth et al. [43] examined catalysts of composition 4Fe/Ni/SiO$_2$ using TPR in association with other techniques such as XRD and Mossbauer spectroscopy. In TPR they found two reduction processes, one a shoulder at 593 K (termed $\propto$) and the other a distinct peak at 678 K (β). When compared with the profiles for Fe/SiO$_2$ and Ni/SiO$_2$ it was clear that the profile for the bimetallic was indicative of the reduction of a single phase and was strikingly similar to the profile of Fe/SiO$_2$, except that the β peak was shifted to lower temperatures. It was concluded that Ni is alloyed with Fe and accelerates the reduction of the Fe owing to the comparative ease with which Ni dissociates molecular hydrogen.

3. Nickel Modified by Sodium and Boron

Lemaitre and Houalla [88] examined 10%w Ni/SiO$_2$ modified by 0-14 at. % Na and 0-10 at. % B. They found that increasing amounts of modifier led to decreased reducibility of the supported NiO species. Two possible reasons for this were suggested: (1) the modifier resulted in an observed increase of the dispersion of NiO, or (2) there was possible modification of the NiO by embedded modifier ions. The latter explanation seems the more likely since improved dispersion usually leads to increased reducibility.

VI. CONCLUSIONS

In this chapter we have reviewed the applications of TPR to the characterization of bulk oxidic materials and supported materials, both zeolite- and conventionally supported. The work reviewed has been with a few exceptions almost totally concerned with characterization of heterogeneous catalysts, since it is in this area that practically all TPR to date has been carried out.

It has been clearly demonstrated that the relatively recent technique of TPR can give excellent reproducibility of results from laboratory to laboratory when similar conditions of experimentation and catalyst preparation are used. In those cases where different workers have found results at variance with others there has usually been a clear explanation for the difference, usually associated with the different conditions and/or different catalyst preparations used by the workers concerned. The technique has been shown to be extremely sensitive, witness the excellent TPR profiles from catalysts containing only a few tenths of a percent of reducible material.

A number of phenomena can be readily identified using TPR. Examples given above include metal dispersion, compound formation between metal and support, interaction between metallic components to give alloys, identification of cation positions in zeolites, optimization of catalyst preparation, improvement and understanding of regeneration procedures for deactivated catalysts, analysis of metal contents, determination of valency of metallic species, and passivation of metal catalysts.

The technique, while predominantly applied to catalyst characterization, has obvious applications in analytical chemistry and corrosion science. The results on catalysts outlined above make such widened application quite clear.

Perhaps most exciting is the possibility of using TPR as a routine quality control technique in small- and large-scale manufacturing of catalysts and other materials and in refineries and chemical plants for quality control of catalyst regeneration procedures.

REFERENCES

1. F. Mahoney, R. Rudham, and J. V. Summers, *J. Chem. Soc., Faraday I 75*: 314 (1979)

2. B. Wichterlova, Z. Tvaruzkova, and J. Novakova, *J. Chem. Soc., Faraday I, 79*: 1573 (1983)

3. P. A. Jacobs, J-P Linart, H. Nijs, and J. B. Uytterhoeven, *J. Chem. Soc., Faraday I, 73*: 1745 (1977)

4. S. J. Gentry, N. W. Hurst, and A. Jones, *J. Chem. Soc., Faraday I, 75*: 1688 (1979)

5. P. A. Jacobs, M. Tielen, J. P. Linart, J. B. Uytterhoeven, and H. Beyer, *J. Chem. Soc., Faraday I, 72*: 2793 (1976)

6. J. J. Verdonck, P. A. Jacobs, and J. B. Uytterhoeven, *J. Chem. Soc., Commun.*, 181 (1979)

7. J. J. Verdonck, P. A. Jacobs, M. Genet, and G. Poncelet, *J. Chem. Soc., Faraday I, 76*: 403 (1980)

8. R. Brown, M. E. Cooper, and D. A. Whan, *Appl. Catal., 3*: 177 (1982)

9. S. D. Robertson, B. D. McNicol, J. H. de Baas, S. C. Kloet, and J. W. Jenkins, *J. Catal., 37*: 424 (1975)

10. D. A. M. Monti and A. Baker, *J. Catal., 83*: 323 (1983)

11. G. Poncelet, P. Jacobs, F. Delannay, M. Genet, P. Gerard, and A. Herbillon, *Bull. Miner., 102*: 379 (1979)

12. P. A. Jacobs, H. H. Nijs, and G. Poncelet, *J. Catal., 64*: 251 (1980)

13. C. Ghesquiere, J. Lemaitre, and A. J. Herbillon, *Clay Miner., 17*: 217 (1982)

14. S. J. Gentry, N. W. Hurst, and A. Jones, *J. Chem. Soc., Faraday I, 77*: 603 (1981)

15. R. Schoepp and I. Hajal, *C. R. Acad. Sci. Paris, t271*: 1283 (1970)

16. B. D. McNicol and R. T. Short, *J. Electroanal. Chem.*, *81*: 249 (1977)

17. B. D. McNicol and R. T. Short, *J. Electroanal. Chem.*, *92*: 115 (1978)

18. A. Bossi, A. Cattalani, F. Gabrassi, G. Petrini, and L. Zanderighi, *J. Thermal Anal.*, *26*: 81 (1983)

19. T. Mahmood, J. O. Williams, R. Miles, and B. D. McNicol, *J. Catal.*, *72*: 218 (1981)

20. M. J. Fuller and M. E. Warwick, *J. Catal.*, *29*: 441 (1973)

21. U.S. Patent 3,702,294 (1972)

22. M. M. P. Janssen and J. Moolhuysen, *J. Catal.*, *46*: 289 (1977)

23. B. D. McNicol, R. T. Short, and A. C. Chapman, *J. Chem. Soc., Faraday I*, *72*: 2735 (1976)

24. V. B. Hughes and B. D. McNicol, *J. Chem. Soc., Faraday I*, *75*: 2165 (1979)

25. H. C. Yao and M. Shelef, *J. Catal.*, *44*: 392 (1976)

26. H. Bosch, B. J. Kip, J. G. Van Ommen, and P. J. Gellings, *J. Chem. Soc., Faraday I*, *80*: 2479 (1984)

27. F. Roozeboom, M. C. Mittelmeljer-Hazelger, J. A. Moulijn, J. Medema, V. H. J. de Beer, and P. J. Gellings, *J. Phys. Chem.*, *84*: 2783 (1980)

28. D. Monti, A. Reller, and A. Baiker, *J. Catal.*, *93*: 360 (1985)

29. A. Lycourghiotis, C. Defosse, F. Delanney, J. Lemaitre, and B. Delmon, *J. Chem. Soc., Faraday I*, *76*: 1677 (1980)

30. T. Paryjczak, J. Rynkowski, and S. Karski, *J. Chromatog.*, *188*: 254 (1980)

31. H. F. J. Van't Blik, Ph.D. Thesis, Eindhoven, The Netherlands (1984)

32. H. C. Yao, M. Sieg, and H. K. Plummer Jr., *J. Catal.*, *59*: 365 (1979)

33. J. K. Vis, Ph.D. Thesis, Eindhoven, The Netherlands (1984)

34. Z. Wu, *Gaodenj Zuexiao Huaxue Xuebao*, 4: 392 (1983)

35. G. W. C. Kaye and T. H. Laby, *Tables of Physical and Chemical Constants*, 14th ed., Longmans, London (1973)

36. A. R. Bailey, *A Textbook of Metallurgy*, McMillan, London (1967)

37. R. Hasegawa, *Trans. Natl. Res. Inst. Metall.*, *20*: 21 (1978)

38. J. M. Thomas and W. J. Thomas, *Introduction to the Principles of Heterogeneous Catalysis*, Academic Press, London (1967)

39. G. K. Boreskov, *Disc. Faraday Soc.*, *41*: 263 (1966)

40. G. K. Boreskov, *Proc. 5th Intern. Congr. Catal.*, North-Holland/Elsevier, Amsterdam (1973) p. 981

41. M. S. Scurrel, *Ann. Rep. Prog. Chem.*, *70*: 87 (1973)

42. M. Bulens, *Ann. Chim.*, *tI*: 13 (1976)

43. E. E. Unmuth, L. H. Schwartz, and J. B. Butt, *J. Catal.*, *63*: 404 (1980)

44. A. Lycourghiotis and D. Vattis, *React. Kinet, Catal. Lett.*, *18*: 377 (1981)

45. R. Burch, A. R. Flambard, M. A. Day, R. L. Moss, N. D. Parkyns, A. Williams, J. Winterbottom, and A. White, in *Preparation of Catalysts III* (G. Poncelet, P. Grange, and P. A. Jacobs, Eds.), Elsevier, Amsterdam (1983)

46. B. Mile and M. A. Zammitt, University of Nottingham, UK. *SERC Grant No. GR/A/3531/0 Final Report* (1983)

47. T. Parysczak, J. Rynkowski, and V. Krzyzanowski, *React. Kinet. Catal. Lett.*, *21*: 295 (1982)

48. M. Shimokawabe, N. Takezawa, and H. Kobayeshi, *Bull. Chem. Soc. Jpn.*, *56*: 1337 (1983)

49. F. S. Delk and A. Vavere, *J. Catal.*, *85*: 380 (1984)

50. B. D. McNicol, paper presented at Surface Reactivity and Catalysis Group Meeting, Nottingham (1976)

51. P. G. J. Koopman, A. P. G. Kieboom, and H. Van Bekkum, *J. Catal.*, *69*: 172 (1981)

52. G. Blanchard, H. Charcosset, M. T. Chenebaux, and M. Primet, *Proc. 2nd Intern. Symp. on Scientific Bases for Prep. of Heterog Catalysts*, Louvain La Neuve, Belgium (1978)

53. H. C. Yao, S. Japar, and M. Shelef, *J. Catal.*, *50*: 407 (1977)

54. B. D. McNicol, *J. Catal.*, *46*: 438 (1977)

55. A. N. Webb, *J. Catal.*, *39*: 485 (1975)

56. M. F. Johnson and V. M. Le Roy, *J. Catal.*, *35*: 434 (1974)

57. C. Bolivar, H. Charcosset, R. Fretz, M. Primet, L. Tournayan, C. Betizeau, G. Leclercq, and R. Maurel, *J. Catal.*, *39*: 249 (1975)

58. D. J. C. Yates and J. H. Sinfelt, *J. Catal.*, *14*: 182 (1969)

59. P. Arnoldy, E. M. Van Oers, O. S. L. Bruinsma, V. H. J. de Boer, and J. A. Moulijn, *J. Catal.*, *93*: 231 (1985)

60. P. Arnoldy, O. S. L. Bruinsma, and J. A. Moulijn, submitted to *J. Molecular Catal.* (1984)

61. R. Thomas, E. M. Van Oers, V. H. J. de Beer, and J. A. Moulijn, *J. Catal.*, *84*: 275 (1983)

62. R. Thomas, E. M. Van Oers, V. H. J. de Beer, and J. A. Moulijn, *J. Catal.*, *76*: 241 (1982)

63. A. J. Van Roosmalen, D. Koster, and J. C. Mol, *J. Phys. Chem.*, *84*: 3075 (1980)

64. R. Thomas, J. A. Moulijn, J. Medema, and V. H. J. de Beer, *J. Molecular Catal.*, *8*: 161 (1980)

65. B. Parlitz, W. Hanke, R. Fricke, M. Richter, U. Roost, and G. Ohlmann, *J. Catal.*, *94*: 24 (1985)

66. P. Arnoldy and J. A. Moulijn, *J. Catal.*, *93*: 38 (1985)

67. R. J. D. Tilley and B. G. Hyde, *J. Phys. Chem. Solids*, *31*: 1613 (1970)

68. T. Paryjczak, J. Goralski, and K. W. Jozwiak, *React. Kinet. Catal. Lett.*, *16*: 147 (1981)

69. J. W. Jenkins, B. D. McNicol, and S. D. Robertson, *Chem. Technol.*, *7*: 316 (1977)

70. K. Foger and H. Jaeger, *J. Catal.*, *67*: 252 (1981)

71. M. F. Johnson and C. D. Keith, *J. Phys. Chem.*, *67*: 200 (1963)

72. T. Huizinga, J. Van Grondelle, and R. Prins, *Appl. Catal.*, in press (1985)

73. H. Lieske, G. Lietz, H. Spindler, and J. Voelter, *J. Catal.*, *81*: 8 (1983)

74. P. A. Attwood, B. D. McNicol, and R. T. Short, *J. Catal.*, *67*: 287 (1981)

75. T. Huizinga and R. Prins, *J. Phys. Chem.*, *55*: 2156 (1981)

76. C. Bolivar, H. Charcosset, R. Fretz, G. Leclercq, and L. Tournagan, *Proc. First European Symp. on Thermal Anal.*, University of Salford, UK (D. Dollimore, Ed.) Heyden (1976), p. 55

77. N. Wagstaff and R. Prins, *J. Catal.*, *59*: 434 (1979)

78. B. H. Isaacs and E. E. Peterson, *J. Catal.*, *77*: 43 (1982)

79. N. Wagstaff and R. Prins, *J. Catal.*, *67*: 255 (1981)

80. A. C. Faro, M. E. Cooper, D. Garden, and C. Kemball, *J. Chem. Research (M)*, 1114 (1983)

81. F. G. Ciapetta and D. N. Wallace, *Catal. Rev.*, *5*: 67 (1976)

82. J. W. Jenkins, paper presented at Surface Reactivity and Catalysis Group Meeting, Brunel, UK. (1979)

83. F. M. Dautzenberg, J. N. Helle, P. Biloen, and W. M. H. Sachtler, *J. Catal.*, *63*: 119 (1980)

84. R. Burch, *J. Catal.*, *71*: 348 (1981)

85. J. M. Anderson, P. J. Conn, and S. G. Brandenberger, *J. Catal.*, *16*: 404 (1970)

86. T. Paryjczak, J. M. Farbotko, and K. W. Jozwiak, *React. Kinet. Catal. Lett.*, *20*: 227 (1982)

87. S. Karski and J. Paryjczak, *React. Kinet. Catal. Lett.*, *15*: 419 (1980)

88. J. Lemaitre and M. Houalla, *C. R. Acad. Sci. (Paris) C*, *291*: 17 (1980)

5

Trends in Temperature Programming for Solid Characterization

I. CURRENT APPLICATIONS

The last 15 years have seen the inception, development, and eventual acceptance of temperature-programmed reduction as a technique for the effective characterization of solid materials. Along the way, as with all new techniques, there have been many learning experiences. In particular we have learned to appreciate the importance of using consistent experimental conditions allowing comparisons to be made between TPR of different materials. The technique has now reached the stage where real comparisons can be made between work of different groups in different laboratories.

Over the years attempts have been amde to put the technique on a quantitative basis [1] and recently the work of Monti and Baiker [2] among others has provided real advances in this area. It is clear that if people adopt an approach to TPR along the lines suggested by Monti and Baiker that at least semi-quantitative data can be obtained. Nevertheless, the greatest strength of TPR has been in monitoring catalyst preparation and regeneration by detecting differences in reducibility brought about by one or another of different pretreatments, or by changes in preparation/regeneration conditions. The information provided by TPR has usually complemented that found by such other techniques as X-ray diffraction, photoelectron spectroscopy, and electron microscopy.

II. FUTURE APPLICATIONS

We feel that TPR will continue to find increasing use in the above applications and that the cost-effectiveness of the technique and its direct relationship to industrial use of materials such as catalysts will be fully realized. Additionally, we predict further use of this technique in chemical plants as a diagnostic tool for large-scale catalyst manufacture and regeneration.

To date, as is quite clear from the contents of this book, TPR has found application predominantly in the field of heterogeneous catalysis, in particular, for catalyst characterization. This is understandable since the technique was first developed by investigators working in this field and, since heterogeneous catalysis is an area where techniques that would characterize catalysts in conditions approaching those used in reality were much needed. However, there are other areas where TPR might

usefully be applied profitably, e.g., in the characterization of
corrosion layers on materials, engine deposits, or reducible
species in coal ashes, and as a purely analytical tool for evaluating the content of known reducible species in solids.

In terms of the experimental side of the TPR technique,
there are many opportunities for advancement. In the years
since its inception, more or less the same experimental apparatus
and approach have been used, that is, a flowing system at or
about atmospheric pressure with thermal conductivity cell detection. There are of course other detectors that could be used
and their merit would have to be seen in relation to the simplicity and low cost of the thermal conductivity cell. The attach-
ment of a mass spectrometer to a TPR system is an exciting, if
expensive prospect and MS detectors have of course been utilized
in other temperature-programmed reaction studies. With an MS
detector the reduction product could be positively identified directly rather than indirectly as at present.

So far all work on TPR has been carried out at or about
atmospheric pressure. There is in our opinion much to be gained
by carrying out TPR experiments at much higher pressures, or at
least pretreatments of materials could be performed at higher
pressures prior to reducing to atmospheric for the reduction
(most reductions in the plant are carried out at or about atmos-
perhic pressure). In this way we could completely mimic a
catalyst regeneration procedure in situ, for instance, and then
monitor any effects on the TPR profile. Thus, we foresee the
utilization of a high-pressure TPR apparatus as a distinct possibility. Indeed, we are aware that such a development is planned
at the University of Hull, England.

The technique of TPR has been used as a diagnostic tool
for the condition of a catalyst. However, there is no reason the
technique could not be used in a different *reaction* mode, e.g.,
in monitoring the activity of catalysts for reduction reactions
such as hydrogenations. In this way it would be possible to
quickly evaluate the effectiveness of catalysts for such a reaction.

Currently TPR is used in a way such that we can tell, hopefully, from its reduction profile why a particular catalyst was good or bad. In the reaction mode we will see this directly. For example, in the gas phase hydrogenation of benzene, a reducing gas of N_2/H_2 could be used with the reaction product cyclohexane and any other products being trapped out prior to the exit arm of the thermal conductivity cell. This approach has already been used for other reactions such as oxidation, methanation, sulfidation. For example, the techniques of temperature-programmed methanation [3]—and, in particular at British Gas laboratories, stepped-programmed methanation—have been extensively developed and are in routine use for the rapid assessment of activity of catalysts for the methanation reaction [4]. Temperature-programmed oxidation has been very helpful, like TPR, in characterizing catalysts, particularly in evaluating passivation phenomena [5] and characterizing propylene oxidation catalysts [6]. This technique too could prove useful in catalyst regeneration operations, e.g., as a technique to characterize the nature of the coke species in deactivated catalysts [7,8].

Because of the transient nature of temperature-programmed reaction, that is, temperature and surface coverage with reactants varying with time, information can be obtained that is not available from the more conventional steady-state measurements; e.g., average rates of reaction at several different sites are determined in steady state reaction measurements. These sites are of different activities, for example, catalytic metal sites and support sites. With the use of temperature-programming techniques the rates of reaction on the different sites can be separated in much the same way that different reducible species can be distinguished in TPR. A good recent example of this is the work on sulfiding of MoO_3 catalysts by Arnoldy et al. [9], in which temperature-programmed sulfiding was used. The different reactive catalytic sites sulfide at different rates and thus the reactivities of all the Mo species can be detected. An excellent comprehensive review of temperature-programmed reaction (including temperature-programmed desorption) appeared in 1983 [10]. This review summarizes in much greater detail the advantages and disadvantages of temperature-programmed reaction techniques.

The techniques of temperature programming could also benefit from increased rapidity, automation, and integration into one single apparatus of all the possibilities, e.g., reduction, oxidation, sulfidation, methanation, ammoniation, and desorption/adsorption. This has been hampered to some extent by the concentration on the thermal conductivity cell detection systems, which do not lend themselves readily to the broad spectrum of reactions mentioned above because of the time taken to stabilize in switching from one reaction to another. In many respects the simplicity and low cost of the thermal conductivity approach which are considerable advantages have hampered progress in this respect. A more universal detection system is a mass spectrometer, which although more expensive than a thermal conductivity detector, could be considerably more flexible. What is needed is perhaps a greater effort from instrument manufacturers in developing a "black box" capable of meeting the above requirements. Hopefully the next decade will see the advent of such a device. The market is clearly there and in due course we see such a development taking place.

In conclusion, then, the last 10 years have been a tremendous growth in the interest and application of temperature programming and, in particular, temperature-programmed reduction and the technique is clearly here to stay. It is our view that this growth will continue with increasing application being found in areas as yet untouched, with commensurate benefits resulting.

REFERENCES

1. N. W. Hurst, S. J. Gentry, A. Jones, and B. D. McNicol, *Catal. Rev.—Sci. Eng.*, *24* (2): 233 (1982)

2. D. A. Monti and A. Baiker, *J. Catal.*, *83*: 323 (1983)

3. A. Tomita and Y. Tamai, *J. Catal.*, *97*: 293 (1972)

4. D. C. Puxley, I. J. Kitchener, C. Komodoros, and N. Parkyns, in *Preparation of Catalysts III* (G. Poncelet, P. Grange, and P. A. Jacobs, Eds.), Elsevier, Amsterdam (1983)

5. J. K. Vis, Ph.D. thesis, Leiden, The Netherlands (1984)

6. T. Uda, T. T. Lin, and G. W. Keulks, *J. Catal.*, *62*: 26 (1980)

7. J. N. Parera, N. S. Figoli, and E. M. Traffano, *J. Catal.*, *79*: 481 (1983)

8. R. Bacaud, H. Charcosset, M. Guenin, R. Torrellat-Hidalgo, and L. Tournayan, *Appl. Catal.*, *1*: 81 (1981)

9. P. Arnoldy, J. A. M. van der Heijkant, G. D. de Bok, and J. A. Moulijn, *J. Catal.*, *92*: 35 (1985)

10. J. L. Falconer and J. A. Schwarz, *Catal. Rev.—Sci. Eng.*, *25* (2): 141 (1983)

INDEX